Integrate Lua with C++

Seamlessly integrate Lua scripting to enhance
application flexibility

Wenhuan Li

BIRMINGHAM—MUMBAI

Integrate Lua with C++

Group Product Manager: Kunal Sawant

Publishing Product Manager: Akash Sharma

Senior Editor: Esha Banerjee

Technical Editor: Maran Fernandes

Copy Editor: Safis Editing

Project Coordinator: Deeksha Thakkar

Proofreader: Safis Editing

Indexer: Manju Arasan

Production Designer: Alishon Mendonca

Business Development Executive: Debadrita Chatterjee

Marketing Coordinator: Sonia Chauhan

Production reference: 1131023

Published by Packt Publishing Ltd.

Grosvenor House

11 St Paul's Square

Birmingham

B3 1RB, UK

ISBN 978-1-80512-861-8

www.packtpub.com

To my father, mother, and uncle who gifted me an 8086 when I was 8 years old. To all ordinary people who work hard and stay true to their honors. To all the developers who are constantly learning and crafting.

– Wenhuan Li

Contributors

About the author

Wenhuan Li is a software architect and full stack developer. He started his career in the chip design industry, where he worked on embedded software and development tools. Over the past decade, Wenhuan has primarily focused on mobile applications and backend services and was associated with big tech companies such as Meta, Bloomberg, Booking.com, and McAfee.

About the reviewer

Emilio Panighetti brings over twenty years of experience in software development, with the last decade dedicated to working on embedded platforms and cloud-native applications in modern C++ and Lua for Oracle Communications, which are widely utilized globally. Emilio began his career as a researcher and assistant teacher in college. Later, he co-founded an early internet provider, before moving on to collaborate with various unified communications providers and development houses. In his spare time, he engages in home IoT projects.

For inquiries, you can contact him via email at `emiliosic@me.com`.

Table of Contents

Part 1 – Lua Basics

1

Getting Your C++ Project Lua-Ready   3

2

Lua Fundamentals   17

Part 2 – Calling Lua from C++

3

How to Call Lua from C++ — 35

4

Mapping Lua Types to C++ — 53

5

Working with Lua Tables   73

Part 3 – Calling C++ from Lua

6

How to Call C++ from Lua   93

7

Working with C++ Types   113

8

Abstracting a C++ Type Exporter   129

Part 4 – Advanced Topics

9

10

11

Preface

This book teaches you how to integrate Lua into C++.

Lua is a simple and efficient scripting language that has been widely adopted in the gaming industry and embedded systems. In those two fields, a large portion of the control programs and applications are written in C++.

By bridging Lua and C++ together, you can open more doors and opportunities for yourself.

Who this book is for

This book is for C++ programmers who are willing to learn how to integrate Lua into their projects. This book can also benefit Lua programmers who want to learn C++ by combining exercises on C++ with relevant topics.

A basic understanding of the C++ programming language is required to fully understand and engage with the book. Although a deeper understanding of C++ may make the book easier to read, you can learn the advanced C++ techniques you encounter in this book on your own through online research.

No prior knowledge of the Lua programming language is needed, although some Lua experience might be beneficial. You can refer to the Lua reference manual frequently to learn the Lua programming language as you progress through this book.

What this book covers

Chapter 1, *Getting Your C++ Project Lua-Ready*, explores how to get and compile Lua source code for use in your own project.

Chapter 2, *Lua Fundamentals*, provides an introduction to the Lua programming language.

Chapter 3, *How to Call Lua from C++*, guides you on how to load Lua scripts and call Lua functions from C++.

Chapter 4, *Mapping Lua Types to C++*, delves into the mechanism of passing arguments to Lua functions and retrieving return values in C++.

Chapter 5, *Working with Lua Tables*, trains you on how to work with Lua tables in C++.

Chapter 6, *How to Call C++ from Lua*, breaks down the process of calling C++ code from Lua.

Chapter 7, *Working with C++ Types*, delves into exporting C++ classes to Lua.

Chapter 8, *Abstracting a C++ Type Exporter*, guides you on how to build a template class that can export any C++ class to Lua.

Chapter 9, *Recapping Lua-C++ Communication Mechanisms*, reviews and summarizes the mechanisms to integrate Lua with C++.

Chapter 10, *Managing Resources*, explores some advanced memory management techniques and resource management principles.

Chapter 11, *Multithreading with Lua*, enables you to work with Lua in a multithreaded environment.

To get the most out of this book

This book will build a Lua executor. Each chapter will focus on one area and add one feature group to the executor. The exercises at the end of the chapter, if provided, are a crucial part to prepare for the next chapter. For the more experienced readers among you, this is why some code might not get implemented optimally in the initial chapters. This will help you stay focused on the topics at hand.

This book assumes that you know C++ and are willing to learn about Lua. If you have no intention of being a Lua programmer, you only need to focus on the related Lua concepts rather than Lua coding.

Because this is not a dedicated book to teach either of the programming languages, when you see the usage of a certain API that you are not familiar with, you can get more information from a Lua or a C++ reference manual. You can enrich your learning journey by frequently bridging your knowledge gaps through online research.

Depending on your style of learning, you can either type all the examples yourself to be more hands-on or download the source code from GitHub, quickly skipping through the chapters and reverting to a specific chapter as needed when you start to implement your own project.

If you are using the digital version of this book, we advise you to type the code yourself or access the code from the book's GitHub repository (a link is available in the next section). Doing so will help you avoid any potential errors related to the copying and pasting of code.

Download the example code files

You can download the example code files for this book from GitHub at `https://github.com/PacktPublishing/Integrate-Lua-with-CPP`. If there's an update to the code, it will be updated in the GitHub repository.

We also have other code bundles from our rich catalog of books and videos available at `https://github.com/PacktPublishing/`. Check them out!

Conventions used

There are a number of text conventions used throughout this book.

`Code in text`: Indicates code words in text, folder names, filenames, file extensions, pathnames, user input and program output.

Here is an example: "Then you create a union type with `std::variant`, saying that this union type may be and can only be from the pre-defined Lua mappings."

A block of code is set as follows:

```
void LuaExecutor::setRegistry(const LuaValue &key,
                              const LuaValue &value)
{
    pushValue(key);
    pushValue(value);
    lua_settable(L, LUA_REGISTRYINDEX);
}
```

When we wish to draw your attention to a particular part of a code block, the relevant lines or items are set in bold:

```
int luaNew(lua_State *L)
{
    ....
    if (type == LUA_TNIL)
    {
        ...
        lua_pushcfunction(L, luaDelete);
        lua_setfield(L, -2, "__gc");
    }
    ...
}
```

Bold: Indicates a new term, an important concept, or crucial software. Here is an example: "All systems with graphic user interfaces will also provide an application in which to start a shell. Often, those applications are called a **terminal** or a **console**."

> **Tips or important notes**
> Appear like this.

Get in touch

Feedback from our readers is always welcome.

General feedback: If you have questions about any aspect of this book, email us at `customercare@packtpub.com` and mention the book title in the subject of your message.

Errata: Although we have taken every care to ensure the accuracy of our content, mistakes do happen. If you have found a mistake in this book, we would be grateful if you would report this to us. Please visit `www.packtpub.com/support/errata` and fill in the form.

Piracy: If you come across any illegal copies of our works in any form on the internet, we would be grateful if you would provide us with the location address or website name. Please contact us at `copyright@packt.com` with a link to the material.

If you are interested in becoming an author: If there is a topic that you have expertise in and you are interested in either writing or contributing to a book, please visit `authors.packtpub.com`.

Share Your Thoughts

Once you've read *Integrate Lua with C++*, we'd love to hear your thoughts! Scan the QR code below to go straight to the Amazon review page for this book and share your feedback.

https://packt.link/r/1805128612

Your review is important to us and the tech community and will help us make sure we're delivering excellent quality content.

Download a free PDF copy of this book

Thanks for purchasing this book!

Do you like to read on the go but are unable to carry your print books everywhere?

Is your eBook purchase not compatible with the device of your choice?

Don't worry, now with every Packt book you get a DRM-free PDF version of that book at no cost.

Read anywhere, any place, on any device. Search, copy, and paste code from your favorite technical books directly into your application.

The perks don't stop there, you can get exclusive access to discounts, newsletters, and great free content in your inbox daily

Follow these simple steps to get the benefits:

1.   Scan the QR code or visit the link below

https://packt.link/free-ebook/9781805128618

2.   Submit your proof of purchase

3.   That's it! We'll send your free PDF and other benefits to your email directly

Part 1 – Lua Basics

This part of the book will guide you through the process of getting and compiling Lua from the source code. Then, you will explore some options to set up your C++ projects to use Lua. This will be followed by a brief introduction to the Lua programming language.

Hence, the aim of this part is to prepare you for the real work of integrating Lua into C++. Feel free to skip through some of the topics if you are already familiar with them.

This part comprises the following chapters:

- *Chapter 1, Getting Your C++ Project Lua-Ready*
- *Chapter 2, Lua Fundamentals*

1

Getting Your C++ Project Lua-Ready

Throughout the course of this book, you will learn how to integrate Lua into your C++ projects. Each chapter will be based on the knowledge learned from the previous chapters. This chapter teaches you how to prepare a C++ project in which to integrate Lua and introduces the tools used in this book so that you can understand the examples better. If you already know how to use some of the tools, please feel free to skip those sections. If not, feel free to do a deeper dive after going through this chapter.

In this chapter, we will cover the following topics:

- Compiling the Lua source code
- Building a C++ project with the Lua library
- Building a C++ project with the Lua source code
- Executing a simple Lua script
- Other toolchain options

Technical requirements

To follow this chapter and this book, you will require the following:

- A working C++ compiler, preferably the **GNU C++ Compiler** or **Clang/LLVM**
- A build automation tool, preferably **Make**
- The code editor of your choice
- The Lua source code
- The source code for this chapter: `https://github.com/PacktPublishing/Integrate-Lua-with-CPP/tree/main/Chapter01`

You do not need prior Lua programming knowledge to understand this chapter. If you have any doubts relating to the Lua code examples in this chapter, that is fine; you can read it as C++ code, although there are syntax differences. You will learn Lua as you progress through this book. While it would be beneficial, you do not need to be a Lua expert if your focus is only on the C++ side.

We decided to use open-source compilers and build tools to work with the code examples in this book because they are readily available to everyone and are also the tools of choice in most large-scale projects.

If you are using a Linux or Mac development machine, the GNU C++ Compiler (or Clang/LLVM) and Make should already be installed. If not, please install the versions that are supported by your system. If you are a Windows user, you can go to the last section of this chapter first: *Other toolchain options*.

The build tool used is called Make. In real projects, you might use other build tools. But Make is a fundamental option with no other dependencies, and other build tools share similar ideas, which makes it suitable for the purpose of this book. If you wish to, you can adapt the examples in this book to another build tool of your choice.

You can download the Lua source code from `https://www.lua.org/download.html`. You can select a specific version, but you will most likely wish to use the latest release. You can also clone the source code from Lua's Git repository, here: `https://www.github.com/lua/lua`. However, this is not officially recommended because Lua is stable and compact with infrequent changes.

Compiling the Lua source code

There are many ways to access the Lua language. If you are using Linux, you can install Lua for development with the distribution's package manager. For Windows, you can also find prebuilt binaries. However, since our goal is to integrate Lua into your C++ project, instead of using it as a standalone interpreter, it's best to build from the source code yourself. When studying Lua, this will help you learn more about Lua. For example, in a modern code editor, *Visual Studio Code* included, you can easily check the declaration and implementation of Lua library functions.

In this section, we will focus on compiling Lua from its source code. Unarchive the downloaded Lua source code archive. Most compression tools support this, and the Lua download site also gives instructions. When you have done this, you will see a simple folder structure:

```
lua-5.4.6 % ls
Makefile README doc src
```

We will learn what the preceding code block does in the next section.

The Lua source code package has a typical **POSIX** (think Linux and Unix) folder structure.

- `Makefile` is the root `Makefile` for the package
- `src` subfolder contains the source code
- `doc` subfolder contains the documentation

Introducing shell

In this section, we will learn what a **shell** does. Throughout this book, examples are given and tested using zsh (Z shell) on a POSIX machine. Another popular shell is bash, using which you can also run the examples in this book directly. Even if you use an **Integrated Development Environment** (**IDE**) and adapt the examples manually to your IDE, understanding the basics of shell commands can help you understand the examples better. All IDEs internally use various command line programs to do their work, which is similar to what we will be doing in a shell.

In very basic terms, a shell provides a command-line interface to the system. When you access a shell interactively, it could also be said that you are accessing a terminal. The words shell and terminal are sometimes used interchangeably, although they are technically different things. Thankfully, we do not need to worry about the terminology differences here. All systems with graphic user interfaces will also provide an application in which to start a shell. Often, those applications are called a **terminal** or a **console**.

You can start a shell and try the following commands. To find out which shell you are using, use this:

```
% echo $SHELL
/bin/zsh
```

The output /bin/zsh indicates that the shell in use is the Z shell.

To go to a directory, use the following command:

```
cd ~/Downloads/lua-5.4.6
lua-5.4.6 %
```

cd is the command to change the current working directory. This goes into the Lua source code folder.

And, as you have seen previously, ls is the command to list the contents of a directory:

```
lua-5.4.6 % ls
Makefile README doc src
```

Another thing of importance is the % sign. It signifies a shell prompt, and different shells or user roles may see a different sign. The part before % is the current working directory. The part after % is what command you would type into the terminal.

This section is only meant to be a brief explanation. If you encounter a shell command that you do not know, check it out online.

Building Lua

In a shell terminal, go to the unarchived Lua source code folder and execute make all test. If your toolchain is found to be working, you will have compiled the Lua library and command-line interpreter. Now, let's inspect the files of interest:

```
lua-5.4.6 % ls src/*.a src/lua
src/liblua.a src/lua
```

liblua.a is the Lua library you can link with. lua is the Lua command-line interpreter.

Let's now try the interpreter to see if we can run it successfully:

```
lua-5.4.6 % src/lua
Lua 5.4.6 Copyright (C) 1994-2022 Lua.org, PUC-Rio
> 1+1
2
> os.exit()
```

In the terminal, execute src/lua to start the interactive Lua interpreter. First, enter 1+1: Lua will print back a result of 2. Then enter os.exit() to exit the Lua interpreter.

You have now successfully compiled the Lua library from its source code. Next, we will see how to use it in your project.

Building a C++ project with the Lua library

Building your C++ project with the Lua library has the benefit of not having to include the 100+ Lua source files in your project and your source control system. However, it has some disadvantages as well. For example, if your project needs to support multiple platforms, you will need to maintain multiple pre-compiled libraries. In such a case, building from the Lua source code might be easier.

Creating a project to work with the Lua library

In the previous section, we built the Lua library from the source code. Now, let's extract it to use within our project.

Remember the Makefile in the root of the source code folder? Open it and you will find the two lines shown here:

```
TO_INC= lua.h luaconf.h lualib.h lauxlib.h lua.hpp
TO_LIB= liblua.a
```

These are the header files and the static library you need.

Create a folder for your project. Within it, create an empty source file named `main.cpp`, an empty `Makefile`, and two empty folders named `include` and `lib`. Copy the header files into the `include` folder and the library file into the `lib` folder. The project folder should look like this:

```
project-with-lua-lib % tree
.
├── Makefile
├── lua
│   ├── include
│   │   ├── lauxlib.h
│   │   ├── lua.h
│   │   ├── lua.hpp
│   │   ├── luaconf.h
│   │   └── lualib.h
│   └── lib
│       └── liblua.a
└── main.cpp
```

If you are using Linux, `tree` is a shell program used to print a folder's hierarchy. If you don't have `tree` installed, no need to worry. You can also examine the folder structure inside your preferred IDE.

Writing the C++ code

We will write a simple C++ program to test whether we can link to the Lua library. In C++, you only need to include one header file for Lua: `lua.hpp`.

Write `main.cpp` as shown here:

```cpp
#include <iostream>
#include <lua.hpp>

int main()
{
    lua_State *L = luaL_newstate();
    std::cout << "Lua version number is "
              << lua_version(L)
              << std::endl;
    lua_close(L);
    return 0;
}
```

The preceding source code opens Lua, prints its build number, and then closes Lua.

Writing the Makefile

As part of the first project, writing the `Makefile` is very simple. Let's use this as a chance to learn some more details about the `Makefile`, if you haven't got a good understanding already. For more information, you can check out the official website at `https://www.gnu.org/software/make/`.

Write the `Makefile` as shown in the following code:

```
project-with-lua-lib: main.cpp
    g++ -o project-with-lua-lib main.cpp -Ilua/include \
        -Llua/lib -llua
```

This is a very basic `Makefile`. In a real project, you would need a more complex `Makefile` to make it more flexible. You will see more flexible examples later in this chapter. Here, the focus is simplicity for the first encounter. This initial `Makefile` has the following elements:

- `project-with-lua-lib` is a `Makefile` target. You can have as many targets as needed in a `Makefile`. When you invoke `make` without an explicit target, it will execute the first one defined in the file.

- `main.cpp` is the target's dependency. You can depend on another target or a file. You can have as many dependencies as needed.

- The target invokes a `g++` command to compile `main.cpp` and link it with the Lua library. You will need to use a tab instead of spaces before the target commands.

- `-o project-with-lua-lib` specifies the name of the compiled executable file. You can change `project-with-lua-lib` to any name you want.

- `-Ilua/include` adds `lua/include` to the search path for header files.

- `-Llua/lib` adds `lua/lib` to the linker search path for libraries.

- `-llua` tells the linker to link with the static Lua library: `liblua.a`.

Testing the project

In a terminal, execute `make` to build the project. Then execute `./project-with-lua-lib` to run the compiled project:

```
project-with-lua-lib % make
project-with-lua-lib % ./project-with-lua-lib
Lua version number is 504
```

As shown in the preceding code, the C++ program will execute and print: `Lua version number is 504`.

Congratulations! You have finished your first C++ project with Lua by linking to the pre-compiled Lua library. In the next section, we will explore how to use Lua source code directly to avoid the disadvantages stated at the beginning of this section.

Building a C++ project with the Lua source code

Building your C++ project with the Lua source code has the benefit that it is always compiled as part of your project, and there is no possibility of any surprises arising from compiler incompatibilities.

The major difference from linking with a pre-compiled Lua library is that we will now compile the Lua library from its source code first. It is also better to use the source code package without modifying it or copying only a few selected files into a new folder hierarchy. This will help in the future if you need to use a newer Lua version. In such a case, all you will need to do is to replace the Lua source code folder with the new version.

Creating a project to work with the Lua source code

To create a project using the Lua source code, we need to go back to the Lua source code package:

```
lua-5.4.6 % ls
Makefile README doc src
```

You will need the `src` subfolder.

Create a folder for a new project. Within it, create an empty `main.cpp` and an empty `Makefile`, and copy the `src` subfolder shown in the preceding shell output as the `lua` subfolder in your project folder. The project structure should look as follows:

```
project-with-lua-src % tree
.
├── Makefile
├── lua
│   ├── Makefile
│   ├── lapi.c
│   ├── ...
│   ├── lzio.c
│   └── lzio.h
└── main.cpp
```

Writing the C++ code

You can write the C++ code in exactly the same way that you did when building with the Lua library.

Writing the Makefile

Let's go one step further compared to the last project, and write two targets for the Makefile as follows:

```
project-with-lua-src: main.cpp
    cd lua && make
    g++ -o project-with-lua-src main.cpp -Ilua -Llua -llua

clean:
    rm -f project-with-lua-src
    cd lua && make clean
```

This Makefile first goes into the lua subfolder and then it builds the Lua library. After this, it compiles the C++ code and links it with the Lua library. The lua subfolder is a copy of the src folder from the Lua source code archive. If you accidentally copied the whole archive there, you may see some compile errors.

The Makefile also includes a clean target. This will delete compiled files. Usually, all build systems will have a clean target implemented.

Testing the project

In a terminal, enter make to build the project. Then enter ./project-with-lua-src to execute the compiled project:

```
project-with-lua-src % make
project-with-lua-src % ./project-with-lua-arc
Lua version number is 504
```

The C++ program will execute and print: Lua version number is 504.

Testing the clean target

Since we have implemented a clean target, let's also test that:

```
project-with-lua-src % make clean
rm -f project-with-lua-src
cd lua && make clean
rm -f liblua.a lua luac lapi.o lcode.o lctype.o ldebug.o
ldo.o ldump.o lfunc.o lgc.o llex.o lmem.o lobject.o
lopcodes.o lparser.o lstate.o lstring.o ltable.o ltm.o
lundump.o lvm.o lzio.o lauxlib.o lbaselib.o lcorolib.o
ldblib.o liolib.o lmathlib.o loadlib.o loslib.o lstrlib.o
ltablib.o lutf8lib.o linit.o lua.o luac.o
```

This cleans the working folder by deleting compiler-generated files. In a production-ready project, you would do more by first outputting all intermediate files to a separate folder during building, most likely called `build` or `output`, and exclude that folder from the source control system.

So far, we have learned two ways to integrate Lua with C++ projects. Next, let's learn how to execute real Lua scripts from C++.

Executing a simple Lua script

To execute a Lua script, you can choose to use either the Lua library or the Lua source code. For production projects, I personally recommend using the source code. For learning purposes, either way is fine. We will use Lua source code in the rest of this book.

> **Can you notice this?**
>
> Even if you choose to use the Lua source code, in the `Makefile`, you are first building the Lua source code into the Lua library and then make your project link to the library. Compared to using the Lua library directly, using the Lua source code is just doing one more step in your project. You can focus more on the similarities rather than the differences.

Now, let's look at a more general project structure.

Creating a project

As said, there will be more complex projects in the following chapters. For now, we will explore a more general project structure. We will build and link to Lua in a shared location instead of making a copy for each project.

The following is the structure of the project as shown within its parent folders:

```
% tree
.
├── Chapter01
│   ├── execute-lua-script
│   │   ├── Makefile
│   │   ├── main.cpp
│   │   └── script.lua
│   ├── project-with-lua-lib
│   └── project-with-lua-src
└── lua
    ├── Makefile
    ├── README
    ├── doc
    └── src
```

The two relevant folders for this project are as follows:

- The execute-lua-script folder contains the main project, inside which there is a C++ source file, a Makefile, and a Lua script file
- The lua folder contains the Lua source code package, which is unarchived and used as-is

The other folders shown indicate how the source code of this book is organized – first into chapters and then into projects. Following the exact structure is optional, as long as you can make it work.

Writing the Makefile

We've seen two simple Makefiles in the previous two projects. Let's write a more flexible Makefile as follows:

```
LUA_PATH = ../../lua
CXX = g++
CXXFLAGS = -Wall -Werror
CPPFLAGS = -I ${LUA_PATH}/src
LDFLAGS = -L ${LUA_PATH}/src
EXECUTABLE = executable


all: lua project

lua:
    @cd ${LUA_PATH} && make

project: main.cpp
    $(CXX) $(CXXFLAGS) $(CPPFLAGS) $(LDFLAGS) \
        -o $(EXECUTABLE) main.cpp -llua

clean:
    rm -f $(EXECUTABLE)
```

This Makefile is more flexible and should be good enough to use as a template for study purposes. It differs from the previous ones in the following aspects:

- It defines a few variables at the beginning of the file. The benefit is that in a real-life project, you would need to compile multiple files. This way, you do not need to repeat yourself in each target. This is also clearer to read.
- The default target, conventionally named all, depends on two other targets.
- In the lua target, there is an @ sign before the command. This will stop printing out the command content in the terminal when make executes.

- LUA_PATH is the path to the Lua source code relative to the folder where this Makefile resides.

- CXX is the conventional variable name to define the C++ compiler program. Use CC for the C compiler.

- CXXFLAGS defines the parameters provided to the C++ compiler. Use CFLAGS for the C compiler.

- CPPFLAGS defines the parameters provided to the C preprocessor, and C++ shares the same preprocessor.

- LDFLAGS defines the parameters provided to the linker. In some development systems, you may need to put LDFLAGS after -o $(EXECUTABLE).

With the Makefile ready, let's write the C++ code.

Writing the C++ code

To execute a Lua script, we need some real actions, which we can get by calling some Lua library functions. Write main.cpp as follows:

```cpp
#include <iostream>
#include <lua.hpp>

int main()
{
    lua_State *L = luaL_newstate();
    luaL_openlibs(L);
    if (luaL_loadfile(L, "script.lua") ||
        lua_pcall(L, 0, 0, 0))
    {
        std::cout << "Error: "
                  << lua_tostring(L, -1)
                  << std::endl;
        lua_pop(L, 1);
    }
    lua_close(L);
    return 0;
}
```

The code is doing the following things:

1. Opening Lua and creating a Lua state with luaL_newstate.

2. Opening Lua standard libs with luaL_openlibs.

3. Loading a Lua script named script.lua with luaL_loadfile and executing it with lua_pcall. We will write script.lua soon.

4.   Outputting an error if the script has failed to execute. This is done inside the `if` clause.

5.   Closing Lua with `lua_close`.

The Lua functions used here, as well as the Lua state, will be explained in detail in *Chapter 3*.

Testing the project

If you have created an empty `script.lua`, delete it. Compile and run the project:

```
execute-lua-script % make
execute-lua-script % ./executable
Error: cannot open script.lua: No such file or directory
```

As expected, it says `script.lua` is not found. Don't worry. You will write it next. The important thing to note is that you have finished coding the C++ part of the project and have already compiled the project.

Writing the Lua script

Write `script.lua` as the one-liner shown here:

```
print("Hello C++!")
```

This will print `Hello C++!`.

Execute the project again without recompiling the C++ code:

```
execute-lua-script % ./executable
Hello C++!
```

Congratulations! You have now executed a Lua script from a C++ program and changed what the C++ program does after it has been compiled.

The next section offers some ideas to set up your development environment differently.

Other toolchain options

If you do not have access to a native POSIX system, there are many other toolchain options. Here we have given two examples. Because your development platform may be different and OS updates and situations change, these only serve as some ideas. You can always research online and experiment to get a comfortable setup for yourself.

Using Visual Studio or Xcode

The Lua source code is written in C and does not need other dependencies. You can copy the `src` folder from the Lua source code package into Visual Studio or Xcode, either into your project directly or by configuring it as a Lua project that your main project depends on. Tweak the project settings as you need. This is quite doable.

Whatever IDE you choose to use, remember to check its license to see whether you can use the IDE for your purpose.

Using Cygwin

If you use Windows, you can get Cygwin for a POSIX experience:

1. Download the Cygwin installer from `https://sourceware.org/cygwin/` and run the installer.
2. During package selection, search for the two packages called `make` and `gcc-g++`. Select them for installation.
3. Change all your shell commands and project `Makefiles` related to Lua slightly. You need to build the Linux flavor of the library explicitly. For example, change `cd lua && make` to `cd lua && make linux`. This is because the Lua `Makefile` could not detect Cygwin as a Linux flavor.
4. Open the Cygwin terminal, and you can build and run projects as shown in the examples in this chapter.

Summary

In this chapter, we have learned how to compile the Lua source code, how to link to the Lua library, and how to include the Lua source code directly in your project. Finally, we executed a Lua script from C++ code. By following these steps yourself, you should be comfortable and confident in including Lua in your C++ projects and prepared for more complex work.

In the next chapter, we will learn the basics of the Lua programming language. If you are already familiar with the Lua programming language, feel free to skip *Chapter 2, Lua Fundamentals*. We will come back to the communications between Lua and C++ in *Chapter 3, How to Call Lua from C++*.

2
Lua Fundamentals

In this chapter, we will learn the basics of the Lua programming language. You do not need to be a Lua expert or even write any Lua code if you only work on the C++ side. However, understanding the basics will make you more efficient when integrating Lua into C++.

If you already know Lua programming, you can skip this chapter. If you have not used Lua for a while, you can use this chapter to recap. If you want to learn more about Lua programming, you can get the official Lua book: *Programming in Lua*. If you do not know Lua programming, this chapter is for you. Coming from C++, you can read any Lua code with the brief explanations on Lua code in this book. You can believe in yourself and research online when you need to.

We will talk about the following language features:

- Variables and types
- Control structures

Technical requirements

You will use the interactive Lua interpreter to follow the code examples in this chapter. We have built it from the Lua source code in *Chapter 1*. You can also use a Lua interpreter from another channel, for example, the one installed by your operating system's package manager. Before continuing, make sure you have access to one.

When you see code examples in this chapter, like the following one, you should try the example in an interactive Lua shell:

```
Lua 5.4.6  Copyright (C) 1994-2022 Lua.org, PUC-Rio
> os.exit()
%
```

The first line is what the Lua interpreter outputs when it starts. Use `os.exit()` to quit the interpreter.

You can find the source code for this chapter within the book's GitHub repository: https://github.com/PacktPublishing/Integrate-Lua-with-CPP/tree/main/Chapter02

Variables and types

While you may well know that C++ is a statically-typed language, Lua is a dynamically-typed language. In C++, when you declare a variable, you give it a clear type. In Lua, each value carries its own type, and you do not need to explicitly specify a type. Also, you do not need to define a global variable before referencing it. Although you are encouraged to declare it – or better yet, use local variables. We will learn about local variables later in this chapter.

In Lua, there are eight basic types: **nil**, **boolean**, **number**, **string**, **userdata**, **function**, **thread**, and **table**.

We will learn about six of these in this chapter: nil, boolean, number, string, table, and function. Let's try some before we go into the details:

```
Lua 5.4.6 Copyright (C) 1994-2022 Lua.org, PUC-Rio
> a
nil
> type(a)
nil
> a = true
> type(a)
boolean
> a = 6
> type(a)
number
> a = "bonjour"
> type(a)
string
```

What happens here is as follows:

1. Interactively, type a to check the value of the global variable a. Since it's not defined yet, its value is nil.

2. Check the type of variable a with type(a). The value is nil in this case because it is not defined.

3. Assign true to a. Use = for assignment; the same as in C++.

4. Now, its type is boolean.

5. Assign 6 to a.

6. Now, its type is number.

7. Assign "bonjour" to a.

8. Now, its type is string.

Every line executed there is also a Lua statement. Unlike a C++ statement, you do not need to put a semicolon at the end of it.

Next, we will learn more about some of the types.

Nil

The `nil` type has only one value to represent a non-value: `nil`. Its meaning is similar to that of **nullptr** (or **NULL**) in C++.

Booleans

The `boolean` type has two values. They are *true* and *false*. It does what **bool** does in C++.

Numbers

The `number` type covers C++'s **int**, **float**, and **double**, and their variants (such as **long**).

Here, we will also learn about Lua arithmetic operators and relational operators, as these are mostly used on numbers.

Arithmetic operators

An arithmetic operator is an operator that performs arithmetic operations on numbers. Lua supports *six* arithmetic operators:

- `+`: Addition
- `-`: Subtraction
- `*`: Multiplication
- `/`: Division
- `%`: Modulus
- `//`: Floor Division

C++ has seven arithmetic operators: +, -, *, /, %, ++ and --. Lua's arithmetic operators are similar to their C++ counterparts, except for the following differences:

1   There is no ++ or -- operator.

2.   C++ does not have the // operator, which returns the integer part of the result after division. C++ can achieve the same implicitly with normal division because C++ is strongly-typed. This we can see in the following example:

```
Lua 5.4.6 Copyright (C) 1994-2022 Lua.org, PUC-Rio
> 5 / 3
1.6666666666667
> 5 // 3
1
```

3. Note that Lua is a dynamically-typed language. This means 5 / 3 does not produce 1 as C++ would do.

Relational operators

A relational operator is an operator that tests some kind of relation between two values. Lua supports *six* relational operators:

- <: Less than

- >: Greater than

- <=: Less than or equal to

- >=: Greater than or equal to

- ==: Equal to

- ~=: Not equal to

The ~= operator tests the negation of equality. This is the same as the ! = operator in C++. The other ones are the same as they are in C++.

Strings

Strings are always a constant in Lua. You cannot change one character in a string and make it represent another string. You create a new string for that.

We can delimit literal strings with double or single quotes. The rest is quite similar to C++ strings, as shown in the following example:

```
Lua 5.4.6 Copyright (C) 1994-2022 Lua.org, PUC-Rio
> a = "Hello\n" .. 'C++'
> a
Hello
C++
> #a
9
```

We have two operators on strings that cannot be found in C++. To concatenate two strings, you can use the .. operator. To check the length of the string, you can use the # operator.

Like the C++ escape sequence, use \ to escape special characters, as you can see in the line break in the preceding output. If you do not want to insert the newline escape sequence, \n, you can use a long string instead.

Long strings

You can use [[and]] to delimit multiple-line strings, for example:

```
a = [[
Hello
C++
]]
```

This defines a string that is equal to the single-line definition `"Hello\nC++\n"`.

Long strings can make strings more readable. You can define **XML** or **JSON** strings easily in Lua source code with long strings.

If your long string contains [[or]] in its content, you can add some equal signs between the opening brackets, for example:

```
a = [=[
x[y[1]]
]=]
b = [==[
x[y[2]]
]==]
```

How many equal signs you add is up to you. However, one is usually enough.

Tables

Lua tables are like **C++ std::map** containers but are more flexible. A table is the only way to construct complex data structures in Lua.

Let's try a Lua table with some actions. In a Lua interpreter, type the following statements one by one and observe the outputs:

```
Lua 5.4.6 Copyright (C) 1994-2022 Lua.org, PUC-Rio
> a = {}
> a['x'] = 100
> a['x']
100
> a.x
100
```

This creates a table with the { } constructor and assigns a value of 100 to the 'x' string key. Another way to construct this table is a = {x = 100}. To initialize a table with more keys, use a = {x = 100, y = 200}.

a.x is an alternative syntax for a['x']. Which one you use is a style preference. But usually, the dot syntax implies that the table is used as a record or in an object-oriented way.

Except for nil, you can use all Lua types as table keys. You can also use different types as values in the same table. You have complete control, as seen next:

```
Lua 5.4.6 Copyright (C) 1994-2022 Lua.org, PUC-Rio
> b = {"Monday", "Tomorrow"}
> b[1]
Monday
> b[2]
Tomorrow
> #b
2

> a = {}
> b[a] = "a table"
> b[a]
a table

> b.a
nil

> #b
2
```

This example explains four points related to the table:

1. It first creates a table in the form of an array. Note that it is indexed starting from 1, not 0. When you give a value-only entry to the table constructor, it will be treated as part of the array. Do you recall that # is the length operator? It can tell the length of the table when it is used to represent a sequence or an array.

2. Then it adds another entry using another table, a, as the key and the "a table" string as the value. This is perfectly fine.

3. Note that b.a is nil because b.a means b['a'] with the 'a' string key, not b[a].

4. Finally, we try to check the length of the table again. We have added 3 entries in the table, but it outputs a length of 2. Coming from C++, this may surprise you: Lua does not provide a built-in way to check table length. The length operator only provides convenience when a table is an array. You are able to use a table as an array and a map at the same time, but you would need to take full responsibility.

Later in this chapter, when we learn about the for control structure, we will find out more about table traversal. Now we will learn about Lua functions.

Functions

Lua functions serve a similar purpose as C++ functions. But unlike in C++, they are also a first-class citizen as one of the basic data types.

We can define a function by doing the following:

1. Start with the `function` keyword.
2. Follow it with a function name and a pair of parentheses, inside which you can define function parameters if needed.
3. Implement the function body.
4. End the function definition with the `end` keyword.

For example, we can define a function as follows:

```
function hello()
    print("Hello C++")
end
```

This will print out `"Hello C++"`.

To define a function to use with the Lua interpreter, you have two choice:

*   In an interactive interpreter, just start to type the function. When you end each line, the interpreter will know, and you can continue to type the next line of the function definition.
*   Alternatively, you can define your functions in another file that can be later imported into the interpreter. This is easier to work on. We will use this way from now on. Try to put your functions in this section in a file named `1-functions.lua`.

To invoke a function, use its name and a pair of parentheses. This is the same as how you invoke a C++ function, for example:

```
Lua 5.4.6 Copyright (C) 1994-2022 Lua.org, PUC-Rio
> dofile("Chapter02/1-functions.lua")
> hello()
Hello C++
```

`dofile()` is the Lua library method to load another Lua script. Here, we load the file where we have defined our Lua functions. If you have changed the script file, you can execute it again to load the latest script.

Next, we will learn about function parameters and function return values.

Function parameters

Function parameters, also called arguments, are values provided to the function when the function is called.

You can define function parameters by providing parameter declarations inside the parentheses after the function name. This is the same as in C++, but you do not need to provide parameter types, for example:

```
function bag(a, b, c)
    print(a)
    print(b)
    print(c)
end
```

When calling the function, you can pass in fewer or more parameters than defined. For example, you can call the bag function we just defined:

```
Lua 5.4.6 Copyright (C) 1994-2022 Lua.org, PUC-Rio
> dofile("Chapter02/1-functions.lua")
> bag(1)
1
nil
nil
> bag(1, 2, 3, 4)
1
2
3
```

You can see what happens when the number of parameters provided is different from the number defined:

- When not enough parameters are passed, the remaining ones will have a nil value.
- When more parameters are passed, the additional ones will be discarded.

You cannot define a default value for function parameters because Lua does not support it at the language level. But you can check in your function, and if a parameter is nil, assign it a default value.

Function results

You can use the return keyword to return function results. It is possible to return multiple values. Let us define two functions that return one and two values, respectively:

```
function square(a)
    return a * a
end
```

```
function sincos(a)
    return math.sin(a), math.cos(a)
end
```

The first function returns the `square` for the given parameter. The second function returns the `sin` and `cos` of the given parameter. Let us give our two functions a try:

```
Lua 5.4.6 Copyright (C) 1994-2022 Lua.org, PUC-Rio
> dofile("Chapter02/1-functions.lua")
> square(2)
4
> sincos(math.pi / 3)
0.86602540378444         0.5
```

You can see from the output that the functions return one and two values, respectively. In this example, `math.pi`, `math.sin`, and `math.cos` are from the Lua `math` library, which is loaded by default in the interactive interpreter. Have you wondered how to make a basic library for our `sincos` function?

Putting a function in a table

From a holistic point of view, the Lua `math` library – and any other library – is just tables holding functions and constant values. You can define your own:

```
Lua 5.4.6 Copyright (C) 1994-2022 Lua.org, PUC-Rio
> dofile("Chapter02/1-functions.lua")
> mathx = {sincos = sincos}
> mathx.sincos(math.pi / 3)
0.86602540378444         0.5
> mathx["sincos"]
function: 0x13ca052d0
> mathx["sincos"](math.pi / 3)
0.86602540378444         0.5
```

We created a table here named `mathx` and assigned our `sincos` function to the `"sincos"` key.

Now you know how to create your own Lua library. To finish our introduction to the Lua types, let us see why we should use local variables.

Local variables and scopes

So far, we have been using global variables because we just reference one when we need to, right? Yes, it is convenient. But the downside is that they will never go out of scope and can be accessed by all functions, related or not. Coming from a C++ background, you will not agree to this.

We can use the **local** keyword to declare a variable as local to the block. A Lua block has a similar concept as a C++ block, within an `if` branch, within a `for` loop, or within a function.

Local variables are good for preventing pollution of the global environment. Try to define two functions to test this:

```
function test_variable_leakage()
    abc_leaked = 3
end

function test_local_variable()
    local abc_local = 4
end
```

In the first function, a local variable is not used, so a global variable named `abc_leaked` will be created. In the second function, a local variable – `abc_local` – is used, which will be unknown outside its function scope. Let us see the effects:

```
Lua 5.4.6 Copyright (C) 1994-2022 Lua.org, PUC-Rio
> dofile("Chapter02/1-functions.lua")
> abc_leaked
nil
> test_variable_leakage()
> abc_leaked
3

> test_local_variable()
> abc_local
nil
```

From the outputs, we can verify the following:

1. First, we try the first function that does not use a local variable. Before calling the function, we verify that there is no global variable named `abc_leaked`. After calling the function, a global variable – `abc_leaked` – is created.

2. Then we try the second function that uses a local variable. No global variable is created in this case.

You should always use local variables when you can. Next, let us familiarize ourselves with Lua's control structures.

Control structures

The Lua control structures are quite similar to the C++ control structures. Try to compare them with their C++ counterparts as you learn about them.

For the code shown in this section, you can put them in another Lua script file named `2-controls.lua`, and import it with `dofile` in the Lua interpreter. You can put each example in a separate function so that you can test the code with different parameters. By now, you should be comfortable using the Lua interpreter, so we will not show how to do it in the rest of this chapter.

We will first explore how to do conditional branching in Lua, and then we will venture into loops.

if then else

The Lua `if` control structure is similar to the C++ one. However, you do not need the parentheses around the test condition, and you do not use the curly brackets. Instead, you will need the `then` keyword and the `end` keyword to delimit the code branches, for example:

```
if a < 0 then a = 0 end

if a > 0 then return a else return 0 end

if a < 0 then
    a = 0
    print("changed")
else
    print("unchanged")
end
```

The `else` branch is optional if you have no operation for that. If you have only one statement in each branch, you can also choose to write everything in one line.

The Lua language design emphasizes simplicity, so the `if` control structure is the only conditional branching control. What if you want to implement something that resembles a C++ **switch** control structure?

Simulating switch

There is no switch control structure in Lua. To simulate it, you can use `elseif`. The following code does that:

```
if day == 1 then
    return "Monday"
elseif day == 2 then
    return "Tuesday"
elseif day == 3 then
    return "Wednesday"
elseif day == 4 then
    return "Thursday"
elseif day == 5 then
    return "Friday"
```

```
elseif day == 6 then
    return "Saturday"
elseif day == 7 then
    return "Sunday"
else
    return nil
end
```

This behaves the same as a C++ `if..else if` control structure. The `if` and `elseif` conditions will be checked one by one until one condition is met and the name of the day of the week is returned.

while

The Lua **while** control structure also resembles the C++ version. You can put a code block between the `do` keyword and the `end` keyword. The following example prints the days of the week:

```
local days = {
    "Monday", "Tuesday", "Wednesday", "Thursday",
    "Friday", "Saturday", "Sunday"
}
local i = 1
while days[i] do
    print(days[i])
    i = i + 1
end
```

We declare a table named `days` and use it as an array. When the `i` index reaches 8, the loop will end because `days[8]` is `nil` and tests as `false`. Coming from C++, you may wonder why we can access the eighth element of a seven-element array. In Lua, there is no index out of bound issue when a table is accessed this way.

You can use `break` to end the loop immediately. This works for the `repeat` loop and the `for` loop as well, which we will explain next.

repeat

The Lua **repeat** does what the C++ `do..while` control structure does, but the ending condition is treated differently. Lua uses the `until` condition to end the loop instead of C++'s `while`.

Let us implement the same code shown for the `while` control structure but with `repeat` this time:

```
local days = {
    "Monday", "Tuesday", "Wednesday", "Thursday",
    "Friday", "Saturday", "Sunday"
}
```

```lua
local i = 0
repeat
    i = i + 1
    print(days[i])
until i == #days
```

#days returns the length of the day array. The code block within `repeat..until` will loop until i reaches this length.

> **Note**
>
> Keep in mind that for a Lua array, the index starts from 1.

To implement the same code with `do..while` in C++, do the following:

```cpp
const std::vector<std::string> days {
    "Monday", "Tuesday", "Wednesday", "Thursday",
    "Friday", "Saturday", "Sunday"
};
size_t i = 0;
do {
    std::cout << days[i] << std::endl;
    i++;
} while (i < days.size());
```

The C++ implementation looks very similar to the Lua version, except for the ending condition as pointed out earlier: `i < days.size()`. We are checking for less than, not equal to.

for, numerical

The numerical **for** loops through a list of numbers. It takes this form:

```lua
for var = exp1, exp2, exp3 do
    do_something
end
```

- var is treated as a local variable scoped to the for block.
- exp1 is the start value.
- exp2 is the end value.
- exp3 is the step and is optional. When not provided, the default step is 1.

To understand this better, let us see an example:

```lua
local days = {
    "Monday", "Tuesday", "Wednesday", "Thursday",
    "Friday", "Saturday", "Sunday"
}
for i = 1, #days, 4 do
    print(i, days[i])
end
```

i is the local variable and has an initial value of 1. The loop will end when i becomes greater than #days. A step of 4 is also provided. So, after each iteration, the effect is i = i + 4. Once you run this code, you will find out that only Monday and Friday are printed.

And, maybe to your surprise, the float type will work as well:

```
Lua 5.4.6  Copyright (C) 1994-2022 Lua.org, PUC-Rio
> for a = 1.0, 4.0, 1.5 do print(a) end
1.0
2.5
4.0
```

As can be verified by the output, the for loop prints from 1.0 and increases the value by 1.5 every time, as long as the value is not greater than 4.0.

for, generic

The generic for loop traverses all values returned by an iterator function. This form of for loops is very convenient for traversing tables.

When we talked about numerical for loops, we saw how they can traverse index-based tables. However, a table in Lua can be more than an array. The most common iterators on tables are pairs and ipairs. They return the key and value pairs for the table. pairs return pairs in an undefined order, as in most hash map implementations. ipairs returns pairs in a sorted order.

Even for an index-based table, if you want to traverse everything, the generic for loop can also be more convenient:

```lua
local days = {
    "Monday", "Tuesday", "Wednesday", "Thursday",
    "Friday", "Saturday", "Sunday"
}
for index, day in pairs(days) do
    print(index, day)
end
```

This loops through the whole array without referring to the array length. The `pairs` iterator returns the key and value pair, one by one, for each loop iteration until all elements in the table are enumerated. After this, the loop ends.

Summary

In this chapter, we have learned about six out of the eight data types in Lua and the four control structures. We have also learned about local variables and why you should use them. This knowledge will prepare you for the rest of the book.

By now, you should be able to read and understand most of the Lua code out there. Some details and topics were intentionally not included in this chapter. You can study more about them when you encounter them.

In the next chapter, we will learn how to call Lua code from C++.

Exercises

1. Locate the standard string manipulation library in the Lua reference manual. Learn about `string.gmatch`, `string.gsub`, and pattern matching. What pattern represents all non-space characters?

2. Using `string.gmatch` and a generic `for` loop, reverse the sentence "C++ loves Lua." The output should be "Lua loves C++."

3. Can you use `string.gsub` and achieve the same with one line of code?

References

The official Lua reference manual: `https://www.lua.org/manual/5.4/`

Part 2 –
Calling Lua from C++

Now that you are familiar with setting up C++ projects with Lua, you will start to learn how to call Lua code from C++.

You will start to implement a general C++ utility class to load and execute Lua code. First, you will learn how to load Lua scripts and call a Lua function. Then, you will explore how to pass arguments to Lua functions and handle return values. Finally, you will dig deeper to master working with Lua tables.

This part comprises the following chapters:

3

How to Call Lua from C++

In this chapter, we will implement a C++ utility class to execute Lua scripts. This serves two purposes. First, by doing this, you will learn in detail how to integrate the Lua library and call Lua code from C++. Second, you will have a Lua wrapper class ready to use. This helps in hiding all the details. We will start with a basic Lua executor and then we will gradually add more features to it as we progress. You will learn about the following:

- Implementing a Lua executor
- Executing a Lua file
- Executing a Lua script
- Understanding the Lua stack
- Operating on global variables
- Calling Lua functions

Technical requirements

Starting from this chapter, we will focus more on code and Lua integration itself and will be brief about toolchain and project settings. However, you can always refer to the book's GitHub repository to get the complete projects. Please make sure you meet the following requirements:

- You need to be able to compile the Lua library from source code. *Chapter 1* covered this.
- You need to be able to write some basic Lua code to test the C++ class that we will write. *Chapter 2* covered this.
- You can create a `Makefile` project, or use an alternative. In *Chapter 1*, we created three `Makefile` projects. We will create a new project for this chapter.
- You can access the source code for this chapter here: `https://github.com/PacktPublishing/Integrate-Lua-with-CPP/tree/main/Chapter03`

Implementing a Lua executor

We will implement a reusable C++ Lua executor class step by step. Let us call it `LuaExecutor`. We will continue to improve this executor by adding new functions to it.

How to include the Lua library in C++ code

To work with the Lua library, you only need three header files:

- `lua.h` for core functions. Everything here has a `lua_` prefix.
- `lauxlib.h` for the auxiliary library (`auxlib`). The auxiliary library provides more helper functions built on top of the core functions in `lua.h`. Everything here has a `luaL_` prefix.
- `lualib.h` for loading and building Lua libraries. For example, the `luaL_openlibs` function opens all standard libraries.

Lua is implemented in C and those three header files are C header files. To work with C++, Lua provides a convenient wrapper, `lua.hpp`, whose content is as follows:

```
extern "C" {
#include "lua.h"
#include "lualib.h"
#include "lauxlib.h"
}
```

In your C++ code, `lua.hpp` is the only Lua header file you need to include. With this sorted out, let us start to work on our Lua executor.

C++ filename extensions

The Lua library uses `hpp` as the header file extension for `lua.hpp`. This is to distinguish it from other C header files from the Lua library. In this book, for our own C++ code, we use `h` for header files with declaration, `hpp` for header files that include all implementation, `cc` for C++ class implementation, and `cpp` for C++ code not part of a class. This is only one way to name the source code files. Feel free to use your own convention.

Getting a Lua instance

We need to get hold of "an instance" of the Lua library to execute Lua scripts. With some other C++ library, you might create a certain class instance and work with the object. With Lua, you get a **Lua state** and pass this state around for different operations.

Our Lua executor will hide the low-level Lua library details and provide a high-level API for your projects. Here is the LuaExecutor.h class definition to start with:

```
#include <lua.hpp>

class LuaExecutor
{
public:
    LuaExecutor();
    virtual ~LuaExecutor();
private:
    lua_State *const L;
};
```

We have defined a constructor, a destructor, and a private member variable of type lua_State, which is the Lua state, and named it L per Lua conventions.

Here is the LuaExecutor.cc class implementation for the definition so far:

```
#include "LuaExecutor.h"

LuaExecutor::LuaExecutor()
    : L(luaL_newstate())
{
    luaL_openlibs(L);
}

LuaExecutor::~LuaExecutor()
{
    lua_close(L);
}
```

The class encapsulates the creation and cleaning up of Lua state:

1. luaL_newstate() creates a new Lua state. We do this in the constructor initializer list.
2. luaL_openlibs(L) opens the Lua standard libraries for the provided Lua state. This makes library functions – for example, string.gmatch – available to be used in Lua scripts.
3. lua_close(L) closes the Lua state and releases its allocated resources – for example, dynamically allocated memory, and so on.

We will now learn more about Lua state.

What is Lua state?

The Lua library maintains no global state, instead, except for `luaL_newstate`, all Lua library functions expect a Lua state as the first parameter. This makes Lua re-entrant and it can be used for multithreaded code with little effort.

Lua state is a structure named `lua_State` that keeps all Lua internal states. To create a Lua state, use `luaL_newstate`. You can use Lua and treat this state transparently, without being concerned about its internal details.

We can compare this to C++ classes. Lua state holds the class member variables and states. Lua library functions serve the purpose of class member functions. To go one step further, consider the C++ **pimpl** (pointer to implementation) idiom:

```
class Lua
{
public:
    void openlibs();
private:
    LuaState *pImpl;
};



struct LuaState
{
    // implementation details
};
```

In this analogy, `class Lua` is our pseudo C++ Lua library; `struct LuaState` is the private implementation that defines and hides the details. In header files, you would only see its forward declaration, not the definition. The `openlibs` public member function uses `pImpl` (the Lua state) internally.

As an advanced topic, C++ member functions will have `this` as the first parameter after being compiled. The Lua library functions expecting `LuaState` as their first parameter can be understood in a similar way: both `this` and `LuaState` point to the private details of the class.

All this information is to make you feel comfortable in passing Lua state around, while at the same time remaining at ease with not operating on it directly. Now, let us go back and continue to build our Lua executor.

Executing a Lua file

In *Chapter 1*, we used the Lua library to load a file and run the script. We will do the same here but in a more proper C++ way. In LuaExecutor.h, add the following new code:

```
#include <string>

class LuaExecutor
{
public:
    void executeFile(const std::string &path);
private:
    void pcall(int nargs = 0, int nresults = 0);
    std::string popString();
};
```

You can of course make all those member functions const, for example, std::string popString() const, because in LuaExecutor we only keep the Lua state L transparently and do not change its value. Here, we are omitting it to prevent too many line breaks in code listings.

executeFile is our public function and the other two are internal helper functions. In LuaExecutor. cc, let us implement executeFile first:

```
#include <iostream>

void LuaExecutor::executeFile(const std::string &path)
{
    if (luaL_loadfile(L, path.c_str()))
    {
        std::cerr << "Failed to prepare file: "
                  << popString() << std::endl;
        return;
    }
    pcall();
}
```

We load the script with luaL_loadfile, providing the file path. luaL_loadfile will load the file, compile it into a **chunk**, and put it onto the **Lua stack**. We will explain what a chunk is and what the Lua stack is soon.

Most Lua library functions will return 0 if they are successful. You can also compare the return value with LUA_OK explicitly, which is defined as 0. In our case, if no error occurs, we will proceed to the next step to call pcall. If there is an error, we will get the error with popString and print it out. Next, implement pcall as follows:

```
void LuaExecutor::pcall(int nargs, int nresults)
{
    if (lua_pcall(L, nargs, nresults, 0))
    {
        std::cerr << "Failed to execute Lua code: "
                  << popString() << std::endl;
    }
}
```

In pcall, we execute the compiled chunk, which is already on the top of the stack, with lua_pcall. This will also remove the chunk from the stack. If an error occurs, we retrieve and print out the error message.

Besides L, lua_pcall takes three more parameters. We are passing 0 for them now. For now, you only need to know that the second parameter is the number of parameters the Lua chunk expects, and the third parameter is the number of values the Lua chunk returns.

Finally, we will implement the last function, popString:

```
std::string LuaExecutor::popString()
{
    std::string result(lua_tostring(L, -1));
    lua_pop(L, 1);
    return result;
}
```

This pops the top of the Lua stack as a string. We will explain more when you learn more about the Lua stack.

We have two concepts to explain before trying out LuaExecutor.

What is a chunk?

A unit of compilation in Lua is called a chunk. Syntactically, a chunk is simply a code block. When put on the stack, a chunk is a value of the function type. So, although not fully accurate, you can consider it a function, and a Lua file an implicit function definition. Further, functions can have embedded functions defined within them.

What is a Lua stack?

A Lua stack is a stack data structure. Each Lua state internally maintains a Lua stack. Each element in the stack can hold a reference to a piece of Lua data. If you recall that a function is also a basic Lua type, you will feel more comfortable about a stack element representing a function. Both Lua code and C++ code can push elements onto and pop elements from the Lua stack, either explicitly or implicitly. We will discuss more about the Lua stack and how our popString function works later in this chapter.

Now you have learned about chunks and the Lua stack, two of the important concepts in Lua, we can test LuaExecutor.

Testing the Lua executor so far

To test our executor, we need to write a Lua script and some C++ test code to call LuaExecutor. Write script.lua as the following one-liner:

```
print("Hello C++")
```

This will print Hello C++ to the console.

Write main.cpp as follows:

```
#include "LuaExecutor.h"
int main()
{
    LuaExecutor lua;
    lua.executeFile("script.lua");
    return 0;
}
```

This will create an instance of LuaExecutor and execute the Lua script file.

Now, write the Makefile:

```
LUA_PATH = ../lua/src
CXX = g++
CXXFLAGS = -Wall -Werror
CPPFLAGS = -I${LUA_PATH}
LDFLAGS = -L${LUA_PATH}
EXECUTABLE = executable

all: lua project

lua:
    @cd ../lua && make
```

```
project: main.cpp LuaExecutor.cc LuaExecutor.h
    $(CXX) $(CXXFLAGS) $(CPPFLAGS) $(LDFLAGS) -o
        $(EXECUTABLE) main.cpp LuaExecutor.cc -llua

clean:
    rm -f $(EXECUTABLE)
```

Compared to the `Makefiles` in *Chapter 1*, we have one more source file, `LuaExecutor.cc`, and one more header file, `LuaExecutor.h`, that the `project` target depends on. Remember to use tabs for indentations, not spaces. You can find explanations on how to write a `Makefile` in *Chapter 1* if you need to revisit it.

With all the test code written, test this out in a terminal:

```
Chapter03 % make
Chapter03 % ./executable
Hello C++
```

If you have done everything correctly, the code will compile. When executed, it will output `Hello C++`, which is from the Lua script file.

We have learned how to execute Lua code from a file. Now let us try to execute a Lua script directly.

Executing a Lua script

In some projects, you may have a file layer abstraction, or you may get a Lua script from a remote server. Then, you cannot pass a file path to the Lua library and ask it to load it for you. You may also want to load the file yourself as a string to do more auditing before executing it. In those situations, you can ask the Lua library to execute a string as a Lua script.

To do this, we will add a new capability to our Lua executor. In `LuaExecutor.h`, add one more function:

```
class LuaExecutor
{
public:
    void execute(const std::string &script);
};
```

This new function will accept the Lua code in a string directly and execute it.

In `LuaExecutor.cc`, add this implementation:

```
void LuaExecutor::execute(const std::string &script)
{
    if (luaL_loadstring(L, script.c_str()))
    {
```

```
        std::cerr << "Failed to prepare script: "
                  << popString() << std::endl;
        return;
    }
    pcall();
}
```

This new function is exactly the same as executeFile with only one difference. It calls the Lua library luaL_loadstring function, which compiles the string as Lua code and puts the compiled chunk onto the stack. Then, pcall will pop and execute the chunk.

Testing it out

Let us test a Lua script. We no longer need a script file now. Write main.cpp as follows:

```
#include "LuaExecutor.h"
int main()
{
    LuaExecutor lua;
    lua.execute("print('Hello Lua')");
    return 0;
}
```

This tells the Lua executor to run the Lua code:

```
print('Hello Lua')
```

Make and run the project and you will see the output Hello Lua.

More on Lua compilation and execution

As explained before, luaL_loadstring and luaL_loadfile will compile the Lua code, while lua_pcall will execute the compiled code. In our LuaExecutor implementation, we are outputting different error messages – Failed to prepare and Failed to execute respectively. Let us test the two different scenarios to understand the execution stages more.

Testing compilation errors

In main.cpp, change the statement to execute the Lua code and make a Lua syntax error intentionally, by deleting the closing parenthesis:

```
lua.execute("print('Hello Lua'");
```

Now recompile the project and run it. You should see the following error output:

```
Failed to prepare script: [string "print('Hello Lua'"]:1:
')' expected near <eof>
```

`pcall` is not called at all, because the Lua code failed to compile.

Testing the runtime error

This time, change the Lua code to the following:

```
lua.execute("print(a / 2)");
```

There is no syntax error. Recompile, run the project, and see the new error:

```
Failed to execute Lua code: [string "print(a / 2)"]:1:
attempt to perform arithmetic on a nil value (global 'a')
```

This is an execution error because the a variable is not defined yet, but we used it for division.

By now, we have a reusable Lua executor that can execute both a Lua script file and Lua code. Let us learn more about the Lua stack before adding even more features to our executor.

Understanding the Lua stack

The Lua stack is used between C/C++ code and Lua code so that they can communicate with each other. Both C++ code and Lua code can operate on this stack either explicitly or implicitly.

We have seen some Lua library functions reading from and writing to the stack. For example, `luaL_loadstring` can push a compiled chuck onto the stack, and `lua_pcall` pops the chunk from the stack. Let us learn some explicit ways to operate on the stack.

Pushing elements

The Lua library provides functions to push different types of values onto the stack:

```
void lua_pushnil     (lua_State *L);
void lua_pushboolean (lua_State *L, int bool);
void lua_pushnumber  (lua_State *L, lua_Number n);
void lua_pushinteger (lua_State *L, lua_Integer n);
void lua_pushstring  (lua_State *L, const char *s);
```

There are more `lua_pushX` functions but the ones shown above are the basic ones. `lua_Number` is a type alias most likely for either `double` or `float`, and `lua_Integer` can be `long`, `long long`, or something else. They depend on how the Lua library is configured and on your operating system defaults. You would need to decide on the scope of the different platforms your project will support and how you would like to map them to C++ types. In most situations, mapping `lua_Number` to `double` and `lua_Integer` to `long` might be good enough, however, if required, you can implement it in a more portable way.

Querying elements

We can use `lua_gettop` to check how many elements are in the stack. The first element in the stack is the bottom of the stack and is indexed with 1. The second element is indexed as 2, and so on. You can also access the stack by referencing the top of the stack. In this referencing system, the top of the stack is indexed with -1, the second from the top is indexed with -2, and so on. You can see the two referencing systems in the following figure:

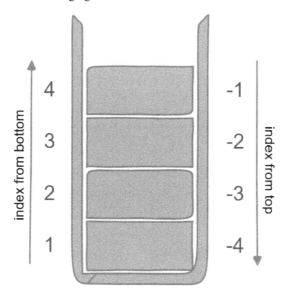

Figure 3.1 – Two ways to access stack elements

As you can see in the figure, each element can be indexed with two numbers. Using the minus number can be very convenient when you need to access the elements you just pushed onto the stack.

Similar to `lua_pushX` for pushing elements, we have `lua_toX` to query elements:

```
int          lua_toboolean (lua_State *L, int index);
const char *lua_tostring   (lua_State *L, int index);
lua_Number  lua_tonumber   (lua_State *L, int index);
lua_Integer lua_tointeger  (lua_State *L, int index);
```

The querying functions will always convert the values to the requested type. This may not be what you want. In this case, you can use `lua_type` to query the type of the element in a given index. There are also corresponding `lua_isX` functions to check whether a given stack index contains a certain type.

Popping elements

To remove the top n elements from the stack, use `lua_pop`:

```
void lua_pop (lua_State *L, int n);
```

For high-level operations in your projects, you should maintain a balanced stack in most situations. This means after you are done, the stack size remains the same as the size before you started. Compared with when you started, if you remove more elements from the stack, you will corrupt the stack and cause undefined behaviors the next time you call Lua. On the other hand, if you remove fewer elements from the stack, you will waste stack space, as well as causing memory leaks. So, popping elements correctly is important at the end of the operation. For example, in our `LuaExecutor::pcall` function, if there is an error, the Lua library will push the error message onto the stack. Because this is triggered by our action, we need to remove the error message with `LuaExecutor::popString`:

```
std::string LuaExecutor::popString()
{
    std::string result(lua_tostring(L, -1));
    lua_pop(L, 1);
    return result;
}
```

This function first reads the top of the stack as a string and then pops the top of the stack.

All communications between C++ and Lua need to use the Lua stack. With a good understanding of the Lua stack, we can continue to learn about Lua global variables.

Operating on global variables

Lua global variables are accessible for the whole Lua state. Consider this Lua code:

```
whom = "C++"
function hello()
    print("Hello " .. whom)
end
```

The `hello` function uses the global variable whom to print out a greeting.

How do we get and set this Lua global variable from C++? We will now extend `LuaExecutor` to do this and use the `hello` function to test it. In this chapter, we will only implement the method to work with string variables to focus primarily on the mechanism.

Getting global variables

You use the Lua library `lua_getglobal` function to get global variables. Its prototype is as follows:

```
int lua_getglobal (lua_State *L, const char *name);
```

`lua_getglobal` expects two parameters. The first one is the Lua state. The second one is the name for the global variable. `lua_getglobal` pushes the value of the global variable onto the stack and returns its type. The types are defined as follows:

```
#define LUA_TNIL           0
#define LUA_TBOOLEAN       1
#define LUA_TLIGHTUSERDATA 2
#define LUA_TNUMBER        3
#define LUA_TSTRING        4
#define LUA_TTABLE         5
#define LUA_TFUNCTION      6
#define LUA_TUSERDATA      7
#define LUA_TTHREAD        8
```

You can check the returned type against those constants to see if the type of the returned data is what was expected.

Let us extend `LuaExecutor` to get global variables. In `LuaExecutor.h`, add a new function declaration:

```
class LuaExecutor
{
public:
    std::string getGlobalString(const std::string &name);
};
```

This function will get a Lua global variable and return it as a string. Implement it in `LuaExecutor.cc`:

```
std::string
LuaExecutor::getGlobalString(const std::string &name)
{
    const int type = lua_getglobal(L, name.c_str());
    assert(LUA_TSTRING == type);
    return popString();
}
```

We call `lua_getglobal` to get the global variable and check to make sure it is of the string type. Then we pop it from the stack with the `popString` function that we implemented earlier to get the Lua library error message.

Setting global variables

To set a Lua global variable from C++, we also utilize the stack. This time, we push the value onto the stack. The Lua library pops it and assigns it to the variable. The Lua library `lua_setglobal` function does the popping and assigning part.

We will add the capability to set global variables to our executor. In `LuaExecutor.h`, add one more function:

```
class LuaExecutor
{
public:
    void setGlobal(const std::string &name,
                   const std::string &value);
};
```

It will set a Lua global variable. The variable's name is provided by the `name` parameter, and the value is set by `value`. In `LuaExecutor.cc`, add the implementation:

```
void LuaExecutor::setGlobal(const std::string &name,
                            const std::string &value)
{
    lua_pushstring(L, value.c_str());
    lua_setglobal(L, name.c_str());
}
```

The code affects the Lua stack as depicted in the following figure:

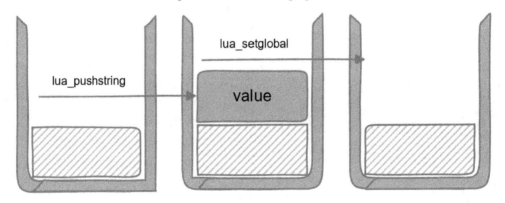

Figure 3.2 – Set global variable

As explained, we first push the value onto the stack with `lua_pushstring`, then invoke the `lua_setglobal` library function to set the global variable. We have maintained a balanced stack size.

Now, let us test our implementations.

Testing it out

We will get and set the whom global variable and call our hello Lua function to test our Lua executor. Rewrite main.cpp as follows:

```cpp
#include <iostream>
#include "LuaExecutor.h"

int main()
{
    LuaExecutor lua;
    lua.executeFile("script.lua");

    std::cout << "Lua variable whom="
              << lua.getGlobalString("whom")
              << std::endl;
    lua.execute("hello()");

    lua.setGlobal("whom", "Lua");

    std::cout << "Lua variable whom="
              << lua.getGlobalString("whom")
              << std::endl;
    lua.execute("hello()");

    return 0;
}
```

The test code is doing four things:

1. Loading script.lua, whose content is the hello function referencing the whom global variable.

2. Calling our getGlobalString executor function to check the value of the whom global variable and executing the Lua hello function to see the truth from Lua's side.

3. Using our setGlobal executor function to change the value of whom to "Lua".

4. Verifying that whom has a new value from both the C++ side and the Lua side.

If you have done everything correctly so far, this test code will output the following:

```
Lua variable whom=C++
Hello C++
Lua variable whom=Lua
Hello Lua
```

Well done for making it this far in the Lua integration journey. With the knowledge of getting and setting global variables, let us move on to the last topic of this chapter: how to call Lua functions from C++.

Calling Lua functions

The Lua `hello` function we used in the previous section is a good example to demonstrate global variables, but it is not how you would usually implement such a feature. Now consider a more suitable implementation:

```
function greetings(whom)
    return "Hello " .. whom
end
```

This Lua `greetings` function expects whom as a function parameter and returns the greeting string instead of printing it out. You can use the greeting string in more flexible ways, for example, by using it on a GUI window.

Earlier in this chapter, while learning how to execute Lua scripts, we implemented the `execute` function in our executor. We can invoke `greetings` with it:

```
LuaExecutor lua;
lua.executeFile("script.lua");
lua.execute("greetings('Lua')");
```

But this is not C++ calling a Lua function; it is a Lua script calling a Lua function. C++ just compiles the Lua script and has no access to the function's return value. To call this function from C++ properly, C++ needs to provide the Lua function parameter and retrieve the return value. By now, this should not be a surprise: you need to use the Lua stack to do that.

Implementing function invocation

We have actually learned everything we need to do this. How it works is a leap of faith in understanding. Let us see the code first and then the explanation.

In `LuaExecutor.h`, add a function to execute a Lua function:

```
class LuaExecutor
{
public:
    std::string call(const std::string &function,
                     const std::string &param);
};
```

This function calls a Lua function, whose name is provided in `function`. It passes a single parameter to the Lua function and expects the Lua function to return a single string type value. It is not very general but is good enough for the purpose of learning for now.

In `LuaExecutor.cc`, implement the `call` function:

```
std::string
LuaExecutor::call(const std::string &function,
                  const std::string &param)
{
    int type = lua_getglobal(L, function.c_str());
    assert(LUA_TFUNCTION == type);
    lua_pushstring(L, param.c_str());
    pcall(1, 1);
    return popString();
}
```

We implemented `pcall` and `popString` earlier in this chapter. The `call` function does the following things:

1. Pushes the Lua function – the name provided in the `function` parameter – onto the stack.

2. Pushes the Lua function parameter – the value provided in the `param` parameter – onto the stack.

3. Calls the Lua `lua_pcall` library function – indicating that the Lua function expects one parameter and returns one value.

Wait! The first line of code looks exactly the same as getting a global variable, no? Indeed, it is! Do you recall that `function` is one of the basic types in Lua? You are getting a global variable, whose name is the function name, and whose value is the function body, onto the stack. Actually, you can also write a Lua function this way:

```
greetings = function (whom)
    return "Hello " .. whom
end
```

This is more cumbersome to write, but shows what is really happening under the hood.

Now, let us look at another similarity:

- In our Lua `execute` and `executeFile` executor functions, we first compile the Lua script as a chunk and put it onto the stack. Then we call `lua_pcall`, indicating a zero count of parameters and a zero count of return values.

- To call a Lua function, we first load the function onto the stack with `lua_getglobal`. Then we push the parameter onto the stack. Finally, we call `lua_pcall` to execute the Lua function, indicating it needs one parameter and will return one value.

Executing Lua scripts is a simplified version of calling a Lua function, without the need to pass parameters and retrieve return values.

By looking at the similarities instead of differences, you will understand better. Now let us test our work.

Testing it out

Rewrite `main.cpp` as follows:

```cpp
#include <iostream>
#include "LuaExecutor.h"

int main()
{
    LuaExecutor lua;
    lua.executeFile("script.lua");
    std::cout << lua.call("greetings", "next adventure")
              << std::endl;
    return 0;
}
```

This will output `"Hello next adventure"` and concludes this chapter.

Summary

In this chapter, we implemented a Lua executor. Not only can it load and execute Lua scripts, but it can also call specific Lua functions. We also learned how to get and set Lua global variables. During the course of the chapter, we explained the Lua stack.

Please take a moment to reflect on how the Lua stack changes during the process of calling a Lua function.

In the next chapter, we will continue to improve this Lua executor and work on Lua data types and C++ data type mappings.

Exercises

1. Implement another function in `LuaExecutor` to call a Lua function with two parameters and two return values. Try to use different Lua data types.

2. In `LuaExecutor`, we are using `std::cerr` to print error messages to the console. So far, the callers cannot get the error state. Design an interface to notify failures. You can pass an implementation of this interface in the `LuaExecutor` constructor.

4

Mapping Lua Types to C++

In the previous chapter, we learned how to call a Lua function with a single string argument that returns one string value. In this chapter, we will learn how to call a Lua function with any type and any number of arguments and support multiple return values. To do this, we need to find a convenient way to map Lua types to C++ types. Then, we will build upon this type system to improve our Lua executor step by step. During this process, you will continue to deepen your understanding of the Lua stack and learn how to use some of the modern C++ features to integrate Lua.

In this chapter, we will cover the following topics:

- Mapping Lua types

- Supporting different argument types

- Supporting a variable number of arguments

- Supporting multiple return values

Technical requirements

This chapter is more C++ coding intensive. To better follow this chapter, please make sure you understand the following:

- You are familiar with the modern C++ standards. We will start to use features from **C++11** and **C++17**. If you have only used **C++03**, please take a while to study the new C++ features on your own when you encounter them in this chapter. As a heads-up, we will use **enum class**, **std::variant**, **std::visit**, and **std::initializer_list**.

- You can access the source code for this chapter here: `https://github.com/PacktPublishing/Integrate-Lua-with-CPP/tree/main/Chapter04`.

- You can understand and execute the code in the `begin` folder from the preceding GitHub link. The `begin` folder has some necessary solutions from the last chapter's questions integrated and acts as the starting point for this chapter. We will add new features implemented in this chapter to it.

Mapping Lua types

In *Chapter 2, Lua Fundamentals*, we learned about the Lua types. They are different from C++ types. To use them in C++, we need to do some mapping. In *Chapter 3, How to Call Lua from C++*, we mapped the Lua string to the C++ `std::string`. It is mapped by hardcoding it in our Lua executor.

What if we want to support all possible Lua types, both in the function arguments and in function return values? What if we want to call Lua functions with different numbers of arguments? It is not feasible to create a C++ function for each argument type and argument count combination. That way, our Lua executor would be plagued with hundreds of functions just to call Lua functions!

Fortunately, C++ is powerful with its **object-oriented programming** and its **generic programming**. These two paradigms lead to two different ways in which you can solve problems in C++.

Exploring different mapping options

As said, C++ supports object-oriented programming and generic programming. We can design a type system with either of them.

Using object-oriented types

This approach is probably easier to understand. It has been supported since the birth of C++. We can define an abstract base class that represents all possible types, then inherit it and implement a concrete class for each type.

Besides C++, most programming languages support this method. If you or your team work with multiple programming languages, this method might cause less concept switching in work.

But this approach is also more verbose. You will have other considerations as well. For example, after the mappings are defined, you would want to prevent creating new types that do not exist in Lua. You would have to make base class constructors private and declare a few friends.

Using generic types

This approach depends on a new C++17 feature: *std::variant*. You can define and map a simple C++ class for each Lua type without inheritance. Then you create a union type with `std::variant`, saying that this union type may be and can only be from the pre-defined Lua mappings.

This will result in less code. The less code, the less chance that something will go wrong. Modern programming tends to adopt new paradigms, besides the vanilla object-oriented methodology.

The drawback of this approach is that not all organizations adopt the new C++ standards that fast, which in turn makes them less widely understood.

In this chapter, we will implement that approach. After going through this chapter, if you prefer, you can implement object-oriented types on your own. But before we move on, let us look at the `Makefile` used for this chapter.

Introducing some new Makefile tricks

Before diving into the details, let us go through the Makefile. You can find a copy in the begin folder in the GitHub repository, as follows:

```
LUA_PATH = ../../lua
CXX = g++
CXXFLAGS = -std=c++17 -Wall -Werror
CPPFLAGS = -I${LUA_PATH}/src
LDFLAGS = -L${LUA_PATH}/src
EXECUTABLE = executable
ALL_O = main.o LuaExecutor.o LoggingLuaExecutorListener.o

all: clean lua project

lua:
    @cd ${LUA_PATH} && make

project: ${ALL_O}
    $(CXX) $(CXXFLAGS) $(CPPFLAGS) $(LDFLAGS) -o $(EXECUTABLE)
      ${ALL_O} -llua

clean:
    rm -f ${ALL_O} $(EXECUTABLE)
```

There are only four differences compared to the Makefile used in *Chapter 3*:

- In CXXFLAGS, we require the compiler to compile our code as C++17 by adding -std=c++17. Without this, it will use the default standard, which may be an older C++ version.

- A new variable, ALL_O, defines all the object files that will be produced. Each C++ source file will be compiled into an object file. Remember to add a new object file here when you add a new source file. Otherwise, without the object file produced, the linker cannot find the symbols that should have been in the missing object file, and you will get linker errors.

- The project target now depends on all object files. Make is smart enough to compile the object files from the source files for you automatically by using the corresponding source file as a dependency for the object file.

- The all target has an extra dependency: the clean target. This always cleans the project and rebuilds it. When you write an object file target manually, you can make it depend on multiple header files. When Make does this for you, it cannot tell which header files need to be depended on. So, this is a trick for small projects for study purposes. For more formal projects, you should consider supporting compiling everything correctly without cleaning it first.

If you struggle to understand this `Makefile`, please check the explanation in *Chapter 1* and *Chapter 3*. Better yet, you can research more online. If you have no urgent need to learn about the `Makefile`, it is also perfectly fine just to use it as it is, and be comfortable with it.

> **Remember**
>
> The `Makefile` examples used in this book favor simplicity, rather than production flexibility.

We have distracted ourselves a bit from C++ by explaining some new `Makefile` mechanisms. This will be the last time we go through a `Makefile` in this book. In the following chapters, please refer to the GitHub source code.

The explanation was necessary, in case you get cryptic errors from the C++ compiler and linker. Now we can go back to our focus. We will define some simple C++ structures that map to Lua types. After this, we can use `std::variant` to declare a union type. Having a union type will enable us to pass a value of any type to our C++ functions. Now, let us define the Lua types in C++.

Defining Lua types in C++

The first order of business is how we can define a Lua type in C++. We want a clear definition of Lua types, so things such as `std::string` are no longer unique enough.

Since C++11, we have *enum class* support. We can limit our Lua types with an enum class in C++ as follows:

```
enum class LuaType
{
    nil,
    boolean,
    number,
    string,
};
```

For now, we only support the Lua basic types that can map to simple C++ types. You can put this declaration in a file named `LuaType.hpp` and include it in `LuaExecutor.h` as shown here:

```
#include "LuaType.hpp"
```

We call it `*.hpp` because we will put the type implementations in the header file directly and inline all the functions. This is partly because the implementation classes will be simple, and partly because this is a book, and limiting the number of lines of code is important. You can separate the code into a header file and a source file or name this header file with implementations `LuaType.h`. This depends on convention, as per each company or organization, and there are many ways to do something in C++.

Implementing Lua types in C++

As explained, we will use simple classes without inheritance. Each class will have two fields: a `type` field of `LuaType` we just defined, and a `value` field for the actual data storage in C++.

In `LuaType.hpp`, implement four structures. In C++, a structure is the same as a class, but with public access to its members by default. We use structures conventionally when we want to define data. First, implement `LuaType::nil`:

```
#include <cstddef>

struct LuaNil final
{
    const LuaType type = LuaType::nil;
    const std::nullptr_t value = nullptr;

    static LuaNil make() { return LuaNil(); }

private:
    LuaNil() = default;
};
```

We choose to use `nullptr` to represent a Lua nil value. Its type is `std::nullptr_t`. We also make the constructor private and provide a static function to create new objects.

> **Design patterns**
>
> Here, we used a design pattern – the static factory method with a private constructor. In our implementation, this will prevent creating objects on the heap with new. The C++ structures for Lua types do not provide copy constructors either. This is a design choice – you either fully support passing it around and assigning to it or limit its usage. In this book, we limit its usage purely when interacting with the Lua executor on the C++ stack. If you have another layer above the Lua executor, you need to convert the structures to C++ basic types or your own type. This helps with abstraction.

Similarly, implement `LuaType::boolean`:

```
struct LuaBoolean final
{
    const LuaType type = LuaType::boolean;
    const bool value;

    static LuaBoolean make(const bool value)
    {
        return LuaBoolean(value);
```

```
        }

    private:
        LuaBoolean(const bool value) : value(value) {}
    };
```

The static `make` function accepts a `boolean` value to create an instance. In the private constructor, we use a member initializer list to initialize the `value` member variable.

For `LuaType::number`, we choose to use the C++ `double` type to store the value:

```
    struct LuaNumber final
    {
        const LuaType type = LuaType::number;
        const double value;

        static LuaNumber make(const double value)
        {
            return LuaNumber(value);
        }

    private:
        LuaNumber(const double value) : value(value) {}
    };
```

Lua itself does not distinguish between an integer and a float in the basic *number* type, but if you need to, you can create two C++ types for integers and floats separately. To do this, you can use the Lua `lua_isinteger` library function to check whether the number is an integer. If it is not, it is a double. In this book, we only implement mappings for the basic Lua types. In a gaming system, you might want to enforce using the float type. In an embedded system, you might want to enforce using the integer type. Or, you can support using both in a project. This is easy to achieve by referencing the implementation for `LuaNumber`.

Linking knowledge

In Lua code, you can use the `math.type` library function to check whether a number is an integer or a float.

And finally, for `LuaType::string`, we use `std::string` to store the value:

```
    #include <string>

    struct LuaString final
    {
        const LuaType type = LuaType::string;
```

```
    const std::string value;

    static LuaString make(const std::string &value)
    {
        return LuaString(value);
    }

private:
    LuaString(const std::string &value) : value(value) {}
};
```

This concludes our type implementations. Next is where all the magic happens.

Implementing a union type

We have defined a LuaType enum class to identify Lua types, and structures to represent Lua values of different types. When we want to pass a Lua value around, we need a type to represent them all. Without using a common base class, we can use std::variant. It is a template class taking a list of types as its parameters. Then it can safely represent any of these types in code. To see it in action, add the following to LuaType.hpp:

```
#include <variant>

using LuaValue = std::variant<
    LuaNil, LuaBoolean, LuaNumber, LuaString>;
```

The using keyword creates a type alias, LuaValue. It can represent any of the four specified types in the template parameters.

Working with the union type

If you have not used std::variant before, you may have wondered how can we tell which actual type it is holding. If you pass a value of LuaValue around, you cannot access the type or the value field directly. This is because there is no common base class. At compile time, the compiler does not know what fields are supported by simply looking at the std::variant variable. To do this, we need a small twist. C++17 also provides std::visit to help with this. Let us implement a helper function to get the LuaType from a LuaValue. In LuaType.hpp, add the code as follows:

```
inline LuaType getLuaType(const LuaValue &value)
{
    return std::visit(
        [](const auto &v) { return v.type; },
        value);
}
```

This function is an **inline** function to make the call site more efficient. Moreover, it needs to be an inline function because we are implementing it in the header file directly. Without the `inline` keyword, the function may get included in different source files with the same symbol, which leads to link errors.

`std::visit` takes two arguments. The first is a C++ **callable** and the second is the value to work on. It passes the value to the callable. In the callable, you can access the field directly, because `std::visit` makes the type information available. If you have never come across this concept or usage before, it may take some time to digest. You can consider this callable a **lambda**. If you have used lambda from other programming languages, such as Java, Kotlin, Swift, or Python, the C++ one is very similar. In other programming languages, more often than not, the lambda is the last parameter, which is called a **trailing lambda** and in some cases is easier to read. The best way to learn about C++ lambda is to use it and try to feel comfortable with it until you have fully mastered it.

With this, let us implement another helper function to get the string representation of each type. In `LuaTypp.hpp`, add the following function:

```
inline std::string
getLuaValueString(const LuaValue &value)
{
    switch (getLuaType(value))
    {
    case LuaType::nil:
        return "nil";
    case LuaType::boolean:
        return std::get<LuaBoolean>(value).value
            ? "true" : "false";
    case LuaType::number:
        return std::to_string(
            std::get<LuaNumber>(value).value);
    case LuaType::string:
        return  std::get<LuaString>(value).value;
    }
}
```

This will help us to test our implementations in the rest of this chapter by getting what is stored in a `LuaValue`. You can use `std::get` to get a specific type from a `std::variant` union.

We have talked about `std::variant`, `std::visit`, and `std::get`, but not enough to be a C++ expert in this area. Before moving on, feel free to research more on them.

Next, let us use the Lua mapping we have implemented to make calling Lua functions more flexible. First, we will get rid of the hardcoded `std::string` used for the `call` function in our Lua executor.

Supporting different argument types

In the last chapter, we implemented our C++ function to call a Lua function as follows:

```
std::string call(const std::string &function,
                 const std::string &param);
```

Our goal in this step is to make it more general and we want the following instead:

```
LuaValue call(const std::string &function,
              const LuaValue &param);
```

In fact, go ahead and change this in `LuaExecutor.h`. To make it work, we will implement helper functions to push onto and pop from the Lua stack, with our `LuaValue` C++ type instead of `std::string`. Let us work on pushing onto the stack first.

Pushing onto the stack

In the previous call function, we pushed the `param` argument of the `std::string` type onto the Lua stack with the following:

```
lua_pushstring(L, param.c_str());
```

To support more Lua types, we can implement a `pushValue` method that takes `LuaValue` as an argument and calls different `lua_pushX` Lua library functions based on the `type` field of `LuaValue`.

In `LuaExecutor.h`, add the declaration as follows:

```
class LuaExecutor
{
private:
    void pushValue(const LuaValue &value);
};
```

And in `LuaExecutor.cc`, implement the `pushValue` function:

```
void LuaExecutor::pushValue(const LuaValue &value)
{
    switch (getLuaType(value))
    {
    case LuaType::nil:
        lua_pushnil(L);
        break;
    case LuaType::boolean:
        lua_pushboolean(L,
            std::get<LuaBoolean>(value).value ? 1 : 0);
```

```
        break;
    case LuaType::number:
        lua_pushnumber(L,
            std::get<LuaNumber>(value).value);
        break;
    case LuaType::string:
        lua_pushstring(L,
            std::get<LuaString>(value).value.c_str());
        break;
    }
}
```

Our implementation only uses one `switch` statement on `LuaType`. We implemented `getLuaType` in `LuaType.hpp` earlier in this chapter. In each `case`, we use `std::get` to get the typed value from the `LuaValue` type union. Next, we will look at the popping part.

Popping from the stack

Popping from the Lua stack is a reverse operation of the pushing part. We will get the value from the Lua stack and use the Lua `lua_type` library function to check its Lua type, and then create a C++ `LuaValue` object with a matching `LuaType`.

To make things more modular, we will create two functions:

- `getValue` to convert a Lua stack position to a `LuaValue`

- `popValue` to pop and return the top of the stack

Add the following declarations to `LuaExecutor.h`:

```
class LuaExecutor
{
private:
    LuaValue getValue(int index);
    LuaValue popValue();
};
```

In `LuaExecutor.cc`, let us first implement `getValue`:

```
LuaValue LuaExecutor::getValue(int index)
{
    switch (lua_type(L, index))
    {
    case LUA_TNIL:
        return LuaNil::make();
```

```
        case LUA_TBOOLEAN:
            return LuaBoolean::make(
                lua_toboolean(L, index) == 1);
        case LUA_TNUMBER:
            return LuaNumber::make(
                (double)lua_tonumber(L, index));
        case LUA_TSTRING:
            return LuaString::make(lua_tostring(L, index));
        default:
            return LuaNil::make();
    }
}
```

The code is quite straightforward. First, we check the Lua type of a requested stack location, and then we return a LuaValue accordingly. For the unsupported Lua types, for example, table and function, we just return LuaNil for now. With this, we can implement popValue as follows:

```
LuaValue LuaExecutor::popValue()
{
    auto value = getValue(-1);
    lua_pop(L, 1);
    return value;
}
```

We first call getValue with -1 as the stack location to get the top of the stack. Then we pop the top of the stack.

With the stack operations implemented, we can now implement the new call function by putting the stack operations together.

Putting it together

Take a moment to read the old call function implementation again. It is shown as follows:

```
std::string LuaExecutor::call(
    const std::string &function,
    const std::string &param)
{
    int type = lua_getglobal(L, function.c_str());
    assert(LUA_TFUNCTION == type);
    lua_pushstring(L, param.c_str());
    pcall(1, 1);
    return popString();
}
```

To implement our new `call` function, not much needs to be changed. We only need to replace the two lines of code doing stack operations with the new helper functions we have just implemented. Write the new call function in `LuaExecutor.cc` as follows:

```
LuaValue LuaExecutor::call(
    const std::string &function, const LuaValue &param)
{
    int type = lua_getglobal(L, function.c_str());
    assert(LUA_TFUNCTION == type);
    pushValue(param);
    pcall(1, 1);
    return popValue();
}
```

We have replaced the lines working with `std::string` with new lines working with `LuaValue`.

Since we have a dedicated `getValue` function and `popValue` to covert a raw Lua value to a `LuaValue`, we can take the chance to make `popString` use them as well. Rewrite it as follows:

```
std::string LuaExecutor::popString()
{
    auto result = std::get<LuaString>(popValue());
    return result.value;
}
```

Here, we have gotten rid of the use of Lua library functions in `popString`. It is a good practice to limit the dependency on a third-party library to only a few functions. Another way to think about this is, in a class, internally, there can be low-level functions and high-level functions.

Next, let us test our improved Lua executor.

Testing it out

Because we have used C++17 features to implement `LuaValue`, we will write the test code in modern C++ as well. Write `main.cpp` as follows:

```
int main()
{
    auto listener = std::make_unique<
        LoggingLuaExecutorListener>();
    auto lua = std::make_unique<LuaExecutor>(*listener);
    lua->executeFile("script.lua");

    auto value1 = lua->call(
        "greetings", LuaString::make("C++"));
```

```
    std::cout << getLuaValueString(value1) << std::endl;

    auto value2 = lua->call(
        "greetings", LuaNumber::make(3.14));
    std::cout << getLuaValueString(value2) << std::endl;

    return 0;
}
```

In this test code, we first use `std::unique_ptr` to hold our Lua executor and its listener, and then load the Lua script with the `greetings` Lua function. This Lua function is from the last chapter. The real actions are calling the Lua function twice: first with `LuaString`, then with `LuaNumber`.

Compile and run the test code. If you have done everything correctly, you should see the following output:

```
Hello C++
Hello 3.14
```

If you see compiler or linker errors, do not feel discouraged. It is common to see a few cryptic error messages when building new C++ code, especially when applying new knowledge. Trace the errors and try to correct them. You can also compare with the code in GitHub if you need to.

> **Note**
>
> We have learned a lot so far. Our improved Lua executor can call Lua functions with a more flexible argument, although it is still only accepting one argument. By now, you should be comfortable and confident in using a common C++ type to represent different Lua types. Take a break and reflect, before moving on to further improve our Lua executor to call Lua functions with a variable number of arguments.

Now, let us continue to improve our Lua executor.

Supporting a variable number of arguments

The Lua `function` supports a variable number of arguments. Let us implement one in `script.lua`:

```
function greetings(...)
    local result = "Hello"
    for i, v in ipairs{...} do
        result = result .. " " .. v .. ","
    end
    return result
end
```

This will return a greeting message and include all arguments in the message. The three dots (. . .) indicate that the function takes a variable number of arguments. We can iterate through the arguments with `ipairs`.

How can we support this in C++? For the stack operations, we only need to push more values. The main decision is how we should declare the Lua executor `call` function to accept a variable number of arguments.

Implementing the C++ function

Since C++11, we can use a **variadic function template** to pass a **parameter pack**. A parameter pack is a list of arguments of any size.

In `LuaExecutor.h`, change the `call` function declaration to the following:

```
template <typename... Ts>
LuaValue call(const std::string &function,
              const Ts &...params);
```

`typename... Ts` defines a template parameter pack, and the function takes it as the `params` argument.

Now, let us implement it. Delete the `call` implementation in `LuaExecutor.cc`. Since we are now using templates, we need to put the implementation in the header file. In `LuaExecutor.h`, add the code as follows:

```
template <typename... Ts>
LuaValue LuaExecutor::call(const std::string &function,
                           const Ts &...params)
{
    int type = lua_getglobal(L, function.c_str());
    assert(LUA_TFUNCTION == type);

    for (auto param :
        std::initializer_list<LuaValue>{params...})
    {
        pushValue(param);
    }

    pcall(sizeof...(params), 1);

    return popValue();
}
```

This implementation can be broken down into four steps, separated by empty lines in code:

- It gets the Lua function to call. This has not changed.

- It pushes the C++ function arguments. Here, we choose to make a `std::initializer_list` from the parameter pack and loop through it.

- It calls the Lua function. We use `sizeof...(params)` to get the size of the parameter pack and tell Lua we are sending that many arguments.

- It gets the return value from the Lua function and returns it.

There is more than one way to accomplish *step 2*. You can use a lambda to unpack the parameter pack, and there are even different options to write this lambda. When, in due course, **C++20** gets adopted, you will have even more options. Those are, however, outside the scope of this book. Here, we choose to use a more conventional way to implement, so that it is easier to understand by more people.

Next, let us test whether our implementation works.

Testing it out

In `main.cpp`, replace the lines that call `lua->call` and print the result as follows:

```
auto result = lua->call("greetings",
    LuaString::make("C++"), LuaString::make("Lua"));
std::cout << getLuaValueString(result) << std::endl;
```

In the test code, we have passed two strings to the Lua `greetings` function. Since we support a variable number of arguments, you can pass as many arguments as you want, zero included. You should see an output similar to `Hello C++, Lua,`.

Some more words on our mechanism

By now, we have implemented in our Lua executor a general function to call any Lua functions, and with any number of arguments. Please take a moment to ponder on the following points, which will deepen your understanding:

- *The Lua function being called does not need to be declared as accepting a variable number of arguments, while our C++ function is.* When calling a Lua function from C++, you always need to tell the Lua library how many arguments have been pushed onto the stack.

- *The Lua function does not need to return a value.* You can try to comment out the return statement in the `greetings` function. The C++ side will get a `LuaNil`, because the Lua library guarantees to push the requested number of return values onto the stack, using nil when the Lua function does not return enough values.

- *The Lua function can return more than one value.* We will only get the first value, and the Lua library will discard the rest, because when calling the Lua function, we requested only one return value.

Our current implementation already supports most of the use cases for calling plain Lua functions, except for the last point mentioned above. Next, we will support multiple return values to complete the Lua function call mechanism.

Supporting multiple return values

To work on retrieving multiple return values, let us first make a Lua function that actually does that. In `script.lua`, add the following function:

```
function dump_params(...)
    local results = {}
    for i, v in ipairs{...} do
        results[i] = i .. ": " .. tostring(v) ..
            " [" .. type(v) .. "]"
    end
    return table.unpack(results)
end
```

This will get every argument and print out its type. We first put them in a table, then we unpack the table so that each table entry is returned as a separate value.

Now, we have some decisions to make. We are happy with the current `call` function except for its return value. However, we cannot overload a function in C++ for a different return type. We need to create another function that returns a list of values.

How can we get multiple return values from Lua? Compared with `call`, there are two differences that we need to tackle:

- How can we tell the Lua library that we are expecting a variable number of return values, instead of a fixed number?

- How can we get this variable number of return values in C++?

To tackle the first problem, while calling the Lua library `lua_pcall` function, we can specify a magic number for the number of expected return values: `LUA_MULTRET`. This means that we will take whatever the Lua function returns, without the library discarding extra return values or padding with nil. This magic number is the only special case to specify the number of return values. It is internally defined as `-1` in `lua.h`.

To tackle the second problem, we only need to count how many elements there are in the Lua stack before calling the Lua function, and how many elements there are after calling the Lua function. This is because the Lua library pushes all return values onto the stack, so the new elements in the stack are the return values. We have implemented `popValue` to pop the top of the stack. We need another function to pop more than one value from the stack.

With the two problems solved, let us start to implement.

Implementing the C++ function

In LuaExecutor.h, add the following declarations:

```
class LuaExecutor
{
public:
    template <typename... Ts>
    std::vector<LuaValue> vcall(
        const std::string &function,
        const Ts &...params);
private:
    std::vector<LuaValue> popValues(int n);
};
```

We added another function to call Lua functions. We call it vcall because it returns a std::vector. We also added a popValues helper function to pop the top n elements from the Lua stack.

First, let us implement vcall in LuaExecutor.h:

```
template <typename... Ts>
std::vector<LuaValue> LuaExecutor::vcall(
    const std::string &function, const Ts &...params)
{
    int stackSz = lua_gettop(L);

    int type = lua_getglobal(L, function.c_str());
    assert(LUA_TFUNCTION == type);

    for (auto param :
        std::initializer_list<LuaValue>{params...})
    {
        pushValue(param);
    }

    if (pcall(sizeof...(params), LUA_MULTRET))
    {
        int nresults = lua_gettop(L) - stackSz;
        return popValues(nresults);
    }
    return std::vector<LuaValue>();
}
```

We now have five steps, which are explained as follows:

1. Record the stack size before doing anything else with `lua_gettop`.

2. Get the Lua function onto the stack with `lua_getglobal`.

3. Push all the arguments onto the stack with `pushValue`.

4. Call the Lua function with `pcall` and pass `LUA_MULTRET` to indicate that we will take all the return values from the Lua function. The Lua library will guarantee to pop all elements you pushed in *step 2* and *step 3*.

5. Pop all the return values from the stack and return them with `popValues`. We check the stack size again. The new stack size minus the original stack size stored in `stackSz` is the number of values returned.

Next, we will implement the final piece, the helper function, to pop all return values from the Lua stack. In `LuaExecutor.cc`, add the following code:

```
std::vector<LuaValue> LuaExecutor::popValues(int n)
{
    std::vector<LuaValue> results;
    for (int i = n; i > 0; --i)
    {
        results.push_back(getValue(-i));
    }
    lua_pop(L, n);
    return results;
}
```

Lua pushes the first return value onto the stack, then the second, and so on. So, the top of the stack needs to be stored at the end of the vector. Here, we read the return values in sequence, starting from the middle of the stack and moving toward the top of the stack. `-i` is the `ith` position counting from the top of the stack.

Next, let's test this out.

Testing it out

In `main.cpp`, change the test code as follows:

```
auto results = lua->vcall(
    "dump_params",
    LuaString::make("C++"),
    LuaString::make("Lua"),
```

```
    LuaNumber::make(3.14),
    LuaBoolean::make(true),
    LuaNil::make());
for (auto result : results)
{
    std::cout << getLuaValueString(result) << std::endl;
}
```

We are passing a list of different types of LuaValue (LuaString, LuaNumber, LuaBoolean and LuaNil) to our new function. This will output the following:

```
1: C++ [string]
2: Lua [string]
3: 3.14 [number]
4: true [boolean]
```

Have you observed anything unusual? We have passed five arguments but only got four return values! LuaNil is not printed out. Why? This is because, in dump_params, we used table.unpack to return multiple values. Lua's table.unpack will stop when it sees a nil value. If you move LuaNil::make() to the middle of the list, you will miss more return values. This is expected. This is a Lua thing. Similar to a C++ char* string, it will end when it first sees a NULL character.

Summary

In this chapter, we first explored how to map Lua types to C++ types, with the goal of ease of use in C++ function calls. Then, we learned about a general way to call any Lua functions.

This chapter progressed step by step. You continued to improve the Lua executor. Each step produced a milestone. This, in turn, was based on the work from the last chapter. Going through the following exercises will also give you a chance to recap what you have learned with hands-on coding. We will continue the book with this methodology.

In the next chapter, we will learn how to integrate Lua tables.

Exercises

1. Implement LuaType::function and LuaFunction to cover the Lua function type. Do not worry about the value field in LuaFunction. You can use nullptr. To test it, you need to call a Lua function that returns another function, and in C++, print out that the return value is a function.

2. Implement LuaType::table and LuaTable to cover the Lua table type. Follow the same instructions as for the previous question.

3. In the last chapter, we implemented `getGlobalString` and `setGlobal` to work with Lua global values. Rewrite those two methods to support more types. You can use the new names `getGlobal` and `setGlobal`, and use `LuaValue`.

4. Implement a private `dumpStack` debug function. This function will dump the current Lua stack. You only need to support the currently supported types in `LuaValue`. Insert a call to this function in different places in `LuaExecutor`. This will deepen your understanding of the Lua stack.

5

Working with Lua Tables

In this chapter, we will continue to improve our Lua executor to work with tables. Many of the mechanisms are extensions of the learnings from the previous chapter. You will also learn about **object-oriented programming** (**OOP**) in Lua and how to call Lua object methods. In all, Lua objects are Lua tables by nature.

We will cover the following topics:

- Working with Lua table entries

- Working with Lua arrays

- OOP in Lua

- Working with Lua table functions

Technical requirements

Here are the technical requirements for this chapter:

- You can access the source code for this chapter at https://github.com/ PacktPublishing/Integrate-Lua-with-CPP/tree/main/Chapter05.

- You can understand and execute the code in the begin folder from the preceding GitHub link. If you haven't already done so, please try to do the exercises in the previous chapter on your own, or at least understand the solutions in the begin folder.

- You can understand the Makefile located in GitHub and can build the projects. Alternatively, you can use your own way to build the source code.

Working with Lua table entries

A **table entry** is the key-value pair for a table element. Lua table keys can be of many data types – for example, of function type. For practical reasons, especially when integrating with C++, we only consider string keys and integer keys.

In `script.lua`, add a simple table as follows:

```
position = { x = 0, y = 0 }
```

`position` is indexed by strings. We will learn how to read from and write to it in C++.

Getting a table entry value

Up until now, in C++ code, we have only used one piece of information to locate a value in Lua. Consider how we implemented `LuaExecutor::getGlobal` and `LuaExecutor::call`. To locate a global variable or to call a function, we pass the name of the variable or the function to a Lua library method.

To work with a table entry, we would need two pieces of information – the table and the table entry key. First, we need to locate the table; after that, we can use the entry key to work on the entry value.

The Lua library method to get an entry value is declared as follows:

```
int lua_gettable(lua_State *L, int index);
```

Wait! We analyzed that we would need two pieces of information to locate a table entry, no? How is it possible that `lua_gettable` only takes one meaningful argument, `index`, besides the Lua state, L? Remember the Lua stack? The top of the stack is commonly used to pass additional information. To quote the Lua reference manual, `lua_gettable` does the following:

> *Pushes onto the stack the value* t[k]*, where* t *is the value at the given* index *and* k *is the value on the top of the stack. This function pops the key from the stack, pushing the resulting value in its place.* (https://www.lua.org/manual/5.4/manual.html#lua_gettable)

As explained, the two keys are both located in the Lua stack. As seen in *Figure 5.1*, before the call, the table entry key must be at the top of the stack, while the table can be in any other position in the stack. This is a Lua design decision. Since you may work on the same table from time to time, you can keep the table reference somewhere in the stack to avoid repeatedly pushing it onto the stack for each access:

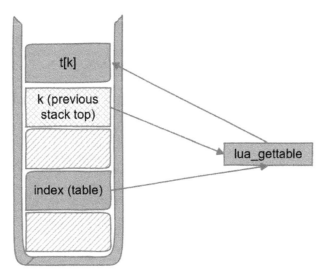

Figure 5.1 – lua_gettable

After understanding the mechanism, it's time to make a design decision. How should we implement the table access in C++? Here are some possibilities:

- We can push the table onto the stack and keep it there. For example, if we are working with a complex table, we can implement a C++ class to load the table at the bottom of the stack and have the C++ object work exclusively with the table.

- We can push the table onto the stack whenever we need to and pop it immediately after it's no longer needed. This works well if the C++ class is working with many Lua values and it is not a performance concern to push the Lua table each time.

Since we are implementing a general Lua executor, we will choose the latter option. In LuaExecutor.h, declare the following function:

```
class LuaExecutor
{
public:
    LuaValue getTable(const std::string &table,
                      const std::string &key);
};
```

It takes the table name and the table entry key name and returns a LuaValue instance. We are only concerned about the string-type key at the moment. In LuaExecutor.cc, implement it as follows:

```
LuaValue LuaExecutor::getTable(
    const std::string &table, const std::string &key)
{
```

```
    int type = lua_getglobal(L, table.c_str());
    assert(LUA_TTABLE == type);
    lua_pushstring(L, key.c_str());
    lua_gettable(L, -2);
    auto value = popValue();
    lua_pop(L, 1);
    return value;
}
```

The code is doing the following things to get a table entry value:

1. It pushes the table reference onto the top of the stack with `lua_getglobal`.

2. It pushes the table entry key onto the top of the stack with `lua_pushstring`. Now, the table is the second from the top.

3. It calls `lua_gettable` to pop the entry key and push the entry value. Now, the entry value is at the top of the stack.

4. It pops the top of the stack as a `LuaValue` with `LuaExecutor::popValue`. Now, the top of the stack is again the table reference.

5. It pops the table with `lua_pop`, as it's no longer needed.

6. It returns the table entry value.

In this implementation, we limit ourselves to only working with tables in the global scope. This is because we are implementing a general Lua executor. For special use cases, you can implement specific C++ classes.

Now, let's see how to set a table entry value.

Setting a table entry value

The Lua library function to set a table entry value is declared as follows:

```
 void lua_settable(lua_State *L, int index);
```

And the quote from the Lua reference manual explains it well:

> *Does the equivalent to* t [k] = v, *where* t *is the value at the given index,* v *is the value on the top of the stack, and* k *is the value just below the top. Pops both the key and the value from the stack. (*`https://www.lua.org/manual/5.4/`
> `manual.html#lua_settable`*)*

This can be seen in *Figure 5.2*. Now, we need to push both the entry key and the entry value onto the Lua stack:

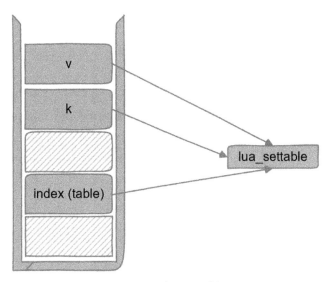

Figure 5.2 – lua_settable

In `LuaExecutor.h`, add the function declaration, as follows:

```
class LuaExecutor
{
public:
    void setTable(const std::string &table,
                  const std::string &key,
                  const LuaValue &value);
};
```

We are passing the value as a `LuaValue`. In `LuaExecutor.cc`, implement it as follows:

```
void LuaExecutor::setTable(const std::string &table,
    const std::string &key, const LuaValue &value)
{
    int type = lua_getglobal(L, table.c_str());
    assert(LUA_TTABLE == type);
    lua_pushstring(L, key.c_str());
    pushValue(value);
    lua_settable(L, -3);
    lua_pop(L, 1);
}
```

The code is explained in the following steps:

1. It pushes the table reference onto the top of the stack with `lua_getglobal`.

2. It pushes the table entry key onto the stack with `lua_pushstring`.

3. It pushes the table entry value onto the stack with `LuaExecutor::pushValue`. Now, the table reference is the third from the top of the stack.

4. It sets the table entry with `lua_settable`. This also pops the top two items from the stack.

5. It pops the table with `lua_pop`. The table is pushed onto the stack in *step 1*.

Next, let's test our implementations so far.

Testing table operations with string keys

In `main.cpp`, add the following helper function to print out a `position` table:

```
void dumpPosition(LuaExecutor *lua)
{
    auto x = lua->getTable("position", "x");
    auto y = lua->getTable("position", "y");
    std::cout << "x=" << std::get<LuaNumber>(x).value
              << ", y=" << std::get<LuaNumber>(y).value
              << std::endl;
}
```

`dumpPosition` calls `LuaExecutor::getTable`, which we have just implemented, to get and print the x field and the y field. In `main()`, change the test code as follows:

```
dumpPositon(lua.get());
lua->setTable("position", "x", LuaNumber::make(3));
lua->setTable("position", "y", LuaNumber::make(4));
dumpPositon(lua.get());
```

This first prints a `position` table, then changes `position.x` to 3 and `position.y` to 4 and prints the table again. If you have done everything correctly, you should see the following output:

```
x=0, y=0
x=3, y=4
```

Next, let's learn how to work with integer types of table keys.

> **Do you remember?**
> If a Lua table uses integer keys exclusively, what else can the table be called?

Working with Lua arrays

Yes – a Lua table with only integer keys is called an array or a sequence. In `script.lua`, add the following array:

```
seq = { 0, 0, 0 }
```

From the C++ side, compared with string keys, the only difference is the data type of the keys. It's straightforward to overload the `getTable` and `setTable` functions by using integer keys. In `LuaExecutor.h`, add the following declarations:

```
class LuaExecutor
{
public:
    LuaValue getTable(const std::string &table,
                      int index);
    void setTable(const std::string &table,
                  int index,
                  const LuaValue &value);
};
```

`index` is the Lua array index – starting from 1. Do not confuse this with the Lua stack index. In the Lua executor's public APIs, there should be no mention of the Lua stack or Lua state.

One way to implement them is to copy the string key version, and instead of calling `lua_pushstring(L, key.c_str())`, call `lua_pushinteger(L, index)`. This will work. But what's the point of repeating ourselves if this is what we would do? Is there another trick?

Using array index optimization

Lua puts lots of emphasis on speed. Because an array is a common form of a Lua table, the Lua library provides special functions to work with arrays, as seen here:

```
void lua_geti(lua_State *L, int index, int key);
void lua_seti(lua_State *L, int index, int key);
```

These functions accept two pieces of information. The `index` argument is the position of the table in the Lua stack. The `key` argument is the array index, as it is also the table entry key. Compared with using `lua_gettable` and `lua_settable`, you no longer need to push the table entry key onto the stack. `lua_seti` expects the value to be at the top of the stack.

Now, let's implement the getTable function for arrays. In LuaExecutor.cc, add the code as follows:

```
LuaValue LuaExecutor::getTable(
    const std::string &table, int index)
{
    int type = lua_getglobal(L, table.c_str());
    assert(LUA_TTABLE == type);
    lua_geti(L, -1, index);
    auto value = popValue();
    lua_pop(L, 1);
    return value;
}
```

The code is doing the following:

1. It gets the table from a global variable and pushes its reference onto the top of the stack.

2. It calls lua_geti with the specified array index. The Lua library will push the value onto the stack.

3. It pops the value as a LuaValue with LuaExecutor::popValue.

4. It pops the table reference.

5. It returns the value.

This does not need to push the array index onto the stack at all. Similarly, implement the setTable function for arrays. In LuaExecutor.cc, add the following code:

```
void LuaExecutor::setTable(const std::string &table,
    int index, const LuaValue &value)
{
    int type = lua_getglobal(L, table.c_str());
    assert(LUA_TTABLE == type);
    pushValue(value);
    lua_seti(L, -2, index);
    lua_pop(L, 1);
}
```

The previous code is explained as follows:

1. It gets the table from a global variable and pushes its reference onto the top of the stack.

2. It pushes the value for the index array position with LuaExecutor::pushValue. Now, the table reference is second from the top of the stack.

3. It calls `lua_seti` to set the array position value. It also pops the value from the stack. Now, the table reference is again at the top of the stack.

4. It pops the table reference.

Next, let's test this.

Testing the array index optimization

In `main.cpp`, add another helper function to print a `seq` Lua array, as follows:

```
void dumpSeq(LuaExecutor *lua)
{
    auto v1 = lua->getTable("seq", 1);
    auto v2 = lua->getTable("seq", 2);
    auto v3 = lua->getTable("seq", 3);
    std::cout << "seq={"
              << std::get<LuaNumber>(v1).value << ", "
              << std::get<LuaNumber>(v2).value << ", "
              << std::get<LuaNumber>(v3).value << "}"
              << std::endl;
}
```

This uses `LuaExecutor::getTable` with integer keys. Replace the test code in `main()` with the following:

```
dumpSeq(lua.get());
lua->setTable("seq", 1, LuaNumber::make(3));
lua->setTable("seq", 2, LuaNumber::make(9));
lua->setTable("seq", 3, LuaNumber::make(27));
dumpSeq(lua.get());
```

This changes the `seq` array to `{ 3, 9, 27 }`. If you have done everything correctly, you should see the output as follows:

```
seq={0, 0, 0}
seq={3, 9, 27}
```

Well done, Lua, for the optimization. And well done, yourself, for making it this far. But how about string keys? In real-world scenarios, more often than not, a Lua table is not an array.

Revisiting string keys

When we first learned to use a string key to access the Lua table, we chose the longer journey to push the key onto the stack. This is because it is a general mechanism, and once learned, you can change to use another data type as table keys.

It is fair to expect an optimization for string keys as well. Here are the Lua library functions for this:

```
int lua_getfield(
    lua_State *L, int index, const char *k);
void lua_setfield(
    lua_State *L, int index, const char *k);
```

These functions work similarly to `lua_geti` and `lua_seti`. `lua_getfield` also returns the type of the table entry value. At the end of this chapter, you will get homework to rewrite `LuaExecutor` with them. You can also choose to do it right now.

Before learning how to call a function from a Lua table, let's write a Lua class. Lua tables with functions are more like C++ objects.

OOP in Lua

OOP in Lua is different than that in C++. In C++, you define a class and create instances of the class. The classes defined are unique types at the language level.

In Lua, there is no native class concept. OOP in Lua is prototype-based. This is more like JavaScript if you are familiar with it. For a Lua table, if an entry is not present, you can instruct Lua to check another table for it, which acts as the prototype for the table you are explicitly referencing.

For ease of understanding, it's fine to call this prototype table the "class" and the table the "object". Or, you can call the relationship "inheritance." Although prototype and class are two different **object-oriented (OO)** methodologies, sometimes people use the two terms interchangeably.

Let's write a class that we will use to learn how to call Lua table functions. Suppose we want to keep a list of places we want to go and note whether we have visited them. In `script.lua`, define a table to be used as the prototype as follows:

```
Destinations = { places = {} }
setmetatable(Destinations.places, {
    __newindex = function (t, k, v)
        print("You cannot use Destinations directly")
    end,
})
```

We defined a table named `Destinations`. It contains a map named `places` to track locations and keep a tab on whether or not they were visited. The key will be the places we want to go, and the value will be *Booleans*. We will define table functions after we have explained how we can use **metatables** to achieve OO behavior.

Using Lua metatables to achieve inheritance

Since Destinations is first of all a plain table, by default you can modify its places entry. How can we prevent users from using it directly? Probably you already know or have guessed. We need to set a Lua metatable. We can use setmetatable to overwrite some operations on the tables. This is comparable to operator overloading in C++.

In our case, we set the __newindex **metamethod** for Destinations.places to a function that does nothing but print an error message. __newindex is called when we assign to an absent table key. This is similar to overloading the C++ subscript operator. We could go to more extremes, but it's fine with this simple limitation to show an attitude.

There is also an __index metamethod we can provide that is used to access absent table keys. This is how we can achieve inheritance behavior. Suppose that we have a table named dst that uses Destinations as its prototype. When we call dst.wish() to add a city to the wish list, what Lua is really doing is first looking up the function via dst["wish"]. Since dst does not have a wish method, Lua calls the __index metamethod, in which we can call the wish method in the Destinations table. This is how Destinations acts as the prototype for dst.

To see it in action, in script.lua, add a constructor for Destinations to create new instances:

```
function Destinations.new(global_name)
    local obj = { places = {} }
    setmetatable(obj, {__index = Destinations})
    if global_name then _G[global_name] = obj end
    return obj
end
```

The new method involves the following steps:

1. It creates a new obj local table with an entry named places, matching the prototype table.

2. It sets obj's __index metamethod as the Destination table. This is another syntax sugar that you can use to set a table as the metamethod. Then, Lua will redirect the lookup for absent keys to the prototype table directly.

3. It assigns the newly created object to a global variable if global_name is provided. Global variables are held in the unique table, _G. If we used the code purely in Lua, we wouldn't need this step. This is to make the new object easily accessible in C++.

4. It returns the new object.

> **More about design decisions**
>
> We provided an option to set a global variable in an object creator. This is a bit unusual and can be considered to cause a side effect from a constructor. You should not blindly copy this paradigm. Consider the following options:
>
> You need to create a Lua executor, do something, and then let it go. This is like invoking a shell command. Most examples in this book use Lua in this way. There is little chance of abusing the global scope. So, assigning the object to a global variable is convenient and efficient.
>
> You need to work with a Lua executor heavily for many things. Then, you can implement a special C++ function to create and keep the table in the stack, and another function to remove it later.
>
> You need to work with a table object exclusively. You might create it in C++'s constructor and keep the table at the bottom of the Lua stack, as pointed out earlier in this chapter.
>
> Better yet, do not use Lua tables at all. In this book, we need to learn how to integrate Lua tables with C++ so that you can do very complex interactions when you need to. But maybe you could divide the C++ domain and the Lua domain more cleanly, and they only send each other simple instructions and results.

With object construction sorted out, we can implement its member functions to make `Destinations` complete.

Implementing Lua class member functions

For a fully functional destination wish list, we need methods to add a place to the wish list. Mark a place visited and check the status of the wish list. Let's define the wish list modifying functions first. In `script.lua`, add the following code:

```
function Destinations:wish(...)
    for _, place in ipairs{...} do
        self.places[place] = false
    end
end

function Destinations:went(...)
    for _, place in ipairs{...} do
        self.places[place] = true
    end
end
```

The `wish` function takes a variable number of arguments, adds them to the `places` map as keys, and sets their values as `false` to indicate an unvisited state. The `went` function is similar and marks its arguments as visited.

The colon operator (:) is a syntax sugar to pass the table as the self parameter to the function. For example, our declaration of the wish function is equivalent to the following:

```
function Destinations.wish(self, ...)
```

Here, self will be the table referenced to call the wish method. This self argument is how most OOP languages work. C++ hides it from you and passes a this pointer to the compiled member methods. Python needs self as the first argument explicitly in member function definitions, with no syntax sugar available. But when calling the Python member functions, you do not need to pass self explicitly.

Next, implement the wish list query functions in script.lua as follows:

```
function Destinations:list_visited()
    local result = {}
    for place, visited in pairs(self.places) do
        if visited then result[#result + 1] = place end
    end
    return table.unpack(result)
end

function Destinations:list_unvisited()
    local result = {}
    for place, visited in pairs(self.places) do
        if not visited then
            result[#result + 1] = place
        end
    end
    return table.unpack(result)
end
```

These functions list visited places and unvisited places respectively.

Testing it out

You can test the Destinations class in a Lua interpreter to make sure it is implemented correctly before using it in C++. Here's an example:

```
Lua 5.4.6  Copyright (C) 1994-2023 Lua.org, PUC-Rio
> dofile("script.lua")
> dst = Destinations.new()
> dst:wish("London", "Paris", "Amsterdam")
> dst:went("Paris")
> dst:list_visited()
Paris
```

```
> dst:list_unvisited()
London   Amsterdam
```

You can add some cities to the wish list, mark one as visited, and print out the list.

With a Lua class ready, we can learn how to call Lua table functions from C++.

Working with Lua table functions

For our Lua executor, we want to call table functions with the same level of support as calling global functions. Similar to `call` and `vcall`, we can define two functions named `tcall` and `vtcall` that call table functions and return a single value and a list of values respectively.

We need to add two more pieces of information to the new C++ member functions – namely, the following:

- The table name, which is obvious
- Whether we should pass the `self` argument to the table function

More on the latter point:

- When the table function does not refer `self` and is used like C++ static member functions, we do not need to pass `self`
- When the table function refers `self` and is used like C++ member functions, we need to pass `self`

Let's implement the code to reinforce what we have just talked about.

Implementing table function support

In `LuaExecutor.h`, add the following declarations:

```
class LuaExecutor
{
public:
    template <typename... Ts>
    LuaValue tcall(
        const std::string &table,
        const std::string &function,
        bool shouldPassSelf,
        const Ts &...params);

    template <typename... Ts>
    std::vector<LuaValue> vtcall(
```

```
        const std::string &table,
        const std::string &function,
        bool shouldPassSelf,
        const Ts &...params);
};
```

`table` is the table name. `function` is the function name, which is a key in the table. `shouldPassSelf` denotes whether we should pass the table as the first argument to the table function. `params` is a list of function arguments.

Next, let's code the `tcall` function in `LuaExecutor.h` as follows; note that the parameter list has been omitted to save space:

```
template <typename... Ts>
LuaValue LuaExecutor::tcall(...)
{
    int type = lua_getglobal(L, table.c_str());
    assert(LUA_TTABLE == type);

    type = lua_getfield(L, -1, function.c_str());
    assert(LUA_TFUNCTION == type);

    if (shouldPassSelf) {
        lua_getglobal(L, table.c_str());
    }

    for (auto param :
        std::initializer_list<LuaValue>{params...}) {
        pushValue(param);
    }

    int nparams = sizeof...(params) +
        (shouldPassSelf ? 1 : 0);
    pcall(nparams, 1);

    auto result = popValue();
    lua_pop(L, 1);
    return result;
}
```

In the previous listing, it's doing six steps separated by newlines, as follows:

1. It gets a global table and pushes it onto the stack.

2. It pushes the table function onto the stack. We are using the `lua_getfield` shortcut.

3. It pushes the table reference onto the stack again, if `shouldPassSelf` is `true`.

4. It pushes the remaining function arguments.

5. It calls the table function. Pay attention to the number of parameters passed.

6. It pops the table function result, pops the table reference pushed in *step 1*, and returns the function result.

If you have done your homework for the previous chapter, you can insert `dumpStack();` at the newlines and see how the Lua stack changes.

Take a moment to digest the `vcall` implementation. And now, you need to implement `vtcall` on your own.

> **Tips**
>
> Reference `vcall` and `tcall`. Pay special attention to getting the count of returned values and where you should put `int stackSz = lua_gettop(L);`.

You can test if you have implemented `vtcall` correctly with the test code that follows.

Testing it out

We will work with the Lua `Destinations` class in C++. In `main.cpp`, replace the test code with the following:

```
lua->tcall("Destinations", "new", false,
    LuaString::make("dst"));
lua->tcall("dst", "wish", true,
    LuaString::make("London"),
    LuaString::make("Paris"),
    LuaString::make("Amsterdam"));
lua->tcall("dst", "went", true,
    LuaString::make("Paris"));
auto visited = lua->vtcall(
    "dst", "list_visited", true);
auto unvisited = lua->vtcall(
    "dst", "list_unvisited", true);
```

This is doing the same thing as when we tested `Destinations` in the interactive Lua interpreter. An explanation for this is provided here:

1. It creates an instance of the class and stores the object in the `dst` global variable. In the `lua->tcall` invocation we set `shouldPassSelf` as `false`.

2. It adds three cities to the wish list of `dst`. From now on, we are working with `dst` and are passing the instance as a `self` argument to the table functions.

3. It marks `Paris` as `visited`.

4. It gets a list of visited cities.

5. It gets a list of unvisited cities.

Add the following lines to print the `visited` and `unvisited` lists:

```
std::cout << "Visited:" << std::endl;
for (auto place : visited) {
    std::cout << std::get<LuaString>(place).value
              << std::endl;
}
std::cout << "Unvisited:" << std::endl;
for (auto place : unvisited) {
    std::cout << std::get<LuaString>(place).value
              << std::endl;
}
```

Compile and run the code. If you have done everything correctly, you should see an output as follows:

```
Visited:
Paris
Unvisited:
London
Amsterdam
```

Congratulations! You have implemented in C++ a mechanism to call Lua table functions. This is by far the most complex logic we have learned!

Summary

In this chapter, we learned how to work with Lua tables in C++. We also touched on OOP in Lua and how it differs from that in C++.

We also explored some design decisions and why `LuaExecutor` is implemented in the way it is. It is designed to learn how to integrate Lua with C++, with a structure that can be broken down into chapters.

By now, you can use `LuaExecutor` to call most Lua scripts, although it has some limitations. For example, we do not support passing another table, except `self`, as a parameter to a function. You can try to implement such a function on your own, but it is likely not a good idea. It is better to keep the communication between Lua and C++ simple.

Take your time to experiment and practice what we have learned. The focus so far is on how to call Lua code from C++. In the next chapter, we will start to learn how to call C++ code from Lua.

Exercises

1. Rewrite the string key overload version of `LuaExecutor::getTable` and `LuaExecutor::setTable`. Use the `lua_getfield` and `lua_setfield` Lua library functions. You can use the same test code in this chapter to test whether you have implemented them correctly.

2. Implement `LuaExecutor::vtcall`. You should have already done so whether you have reached this point.

Part 3 –
Calling C++ from Lua

With your knowledge of calling Lua from C++, in this part, you will continue to learn how to call C++ from Lua.

You will start by learning how to implement and export a C++ function that can be called from Lua scripts. Then, the complexity will increase step by step. You will export a C++ class as a Lua module and improve the process of how it is exported. Finally, you will have a general module exporter that can help you to export any C++ class to Lua.

This part comprises the following chapters:

- *Chapter 6, How to Call C++ from Lua*

- *Chapter 7, Working with C++ Types*

- *Chapter 8, Abstracting a C++ Type Exporter*

6

How to Call C++ from Lua

In the previous three chapters, we focused on learning how to call Lua from C++. In this chapter, we will start to learn how to call C++ from Lua. This is important for your applications because although Lua scripts can extend your C++ applications, they can also benefit from the functions provided by your native C++ code.

This also means we will learn more concepts and piece different things together to make it work. Although the chapters are laid out in a way that they extend the previous chapter in a seamless flow, you may need a different pace to absorb the new concepts. Do read the sections more times if you need practice with coding.

We will cover the following topics:

- How to register C++ functions
- How to override Lua library functions
- How to register C++ modules

Technical requirements

Here are the technical requirements for this chapter:

- You can access the source code for this chapter at `https://github.com/PacktPublishing/Integrate-Lua-with-CPP/tree/main/Chapter06`
- Based on the learnings from the last chapter, you should now be confident in adding code to our Lua executor

This chapter will introduce many new concepts and Lua library APIs. You can cross-check the Lua reference manual online to reinforce the learning: `https://www.lua.org/manual/5.4/`.

How to register C++ functions

Lua is written in C, so it cannot access your C++ classes directly. The only way to call C++ code from Lua is to make it call C++ functions – that is, plain C functions.

How to declare C++ functions for Lua

To register a function to Lua, it must conform to the following prototype:

```
typedef int (*lua_CFunction) (lua_State *L);
```

The function receives only one argument, which is a Lua state. It needs to return an integer value indicating how many return values it produces. The Lua state is private to the function call, and its stack holds the arguments passed from the Lua code when calling the C++ function. The C++ function needs to push its return values onto the stack.

We will first implement a simple function and export it to Lua. Then, we'll see more complex examples to understand more.

Implementing your first C++ function for Lua

Let us add a simple but useful capability to our Lua executor. It will provide a function to check its version code so that the Lua code it executes can query it. In LuaExecutor.cc, right below the #include directives, add the following function implementation that conforms to lua_CFunction:

```
namespace
{
    int luaGetExecutorVersionCode(lua_State *L)
    {
        lua_pushinteger(L, LuaExecutor::versionCode);
        return 1;
    }
}
```

The function pushes a LuaExecutor::versionCode integer constant to its private stack and returns 1 to indicate that it returns one value. We can define this constant in LuaExecutor.h as follows:

```
class LuaExecutor
{
public:
    static const int versionCode = 6;
};
```

We will use the value 6 for *Chapter 6*.

You may have noticed that the function is inside an anonymous namespace. This is to make sure that it cannot be accessed outside the file of LuaExecutor.cc. This also helps with logical code grouping.

Next, let us make this function available to Lua.

How to register C++ functions to Lua

There are a few ways to register C++ functions to Lua. We will look at the simplest way here to register a C++ function in the Lua global table. Later in this chapter when learning C++ modules, we will learn a more proper way to register C++ functions in their own table. You already know how to do this from *Chapter 3*. I only need to point it out with the following code, which you should add to the same anonymous namespace you have just written:

```
namespace
{
    void registerHostFunctions(lua_State *L)
    {
        lua_pushcfunction(L, luaGetExecutorVersionCode);
        lua_setglobal(L, "host_version");
    }
}
```

Yes—we simply need to set it as a Lua global variable! We use lua_pushcfunction to push the lua_CFunction type onto the stack. Then, we assign it to a global variable named host_version.

Using a global variable for the host executor version sounds very reasonable. But you should not abuse Lua global variables by using them too much. Now, let us try it out.

Testing it out

We need to modify three places to test our progress so far. You can start your work with the begin folder from the source code of this chapter.

Call registerHostFunctions from the constructor of our Lua executor, as follows:

```
LuaExecutor::LuaExecutor(...)
{
    ...
    registerHostFunctions(L);
}
```

This registers our function to Lua.

Replace the content of script.lua as follows:

```
print("Host version is " .. host_version())
```

This calls our C++ function from Lua and prints out the result.

Replace the content of `main.cpp` with the following test code:

```cpp
#include "LuaExecutor.h"
#include "LoggingLuaExecutorListener.h"

int main()
{
    auto listener = std::make_unique<
        LoggingLuaExecutorListener>();
    auto lua = std::make_unique<LuaExecutor>(*listener);
    lua->executeFile("script.lua");
    return 0;
}
```

This resets the test code to simply create a Lua executor and runs `scripts.lua`. Run the project, and if you have done everything correctly, you should see the following output:

```
Host version is 6
```

Congratulations! You have called your first C++ function from Lua code. Based on this learning, let us find out how to override Lua library functions.

How to override Lua library functions

Why would you want to override Lua library functions? First, it helps to learn more about calling C++ functions from Lua in a progressive way, before moving on to C++ modules. Second, but more importantly, it is a frequent requirement for real-life projects.

Suppose you are working on a game where assets are packed inside a private archive and your Lua scripts need to access them. Overriding the Lua `io` and `file` libraries can provide a seamless experience for your fellow Lua developers and enforce security at the same time. You can make sure Lua scripts can only access assets you want them to, but nothing else on the host filesystem. This is even more important when your users can change the Lua scripts.

Let us implement a more trivial case. We use the Lua `print` function to output debug information. We want to merge the Lua debug output with C++ output so that we get all our logs in the same place ordered by the time they are printed.

Reimplementing the Lua print function

Because the Lua `print` function takes a variable number of arguments, we need to take this into consideration in our implementation. In `LuaExecutor.cc`, below the namespace from the previous section, add *another* namespace as follows:

```
namespace
{
    int luaPrintOverride(lua_State *L)
    {
        int nArgs = lua_gettop(L);
        std::cout << "[Lua]";
        for (int i = 1; i <= nArgs; i++)
        {
            std::cout << " "
                      << luaL_tolstring(L, i, NULL);
        }
        std::cout << std::endl;
        return 0;
    }
}
```

The `luaPrintOverride` C++ function would eventually get called when you invoke the `print` function in Lua. It takes `lua_State` as a single argument, whose associated Lua stack is used to pass the real arguments from the Lua call site. To understand what is happening, see the following diagram:

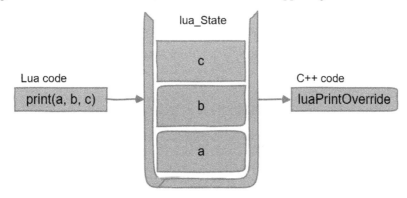

Figure 6.1 – Overriding the Lua print function

The Lua `print` function will push its arguments onto the private Lua stack for the call. The C++ function first checks the number of arguments the Lua call site has passed with `lua_gettop`. Then, it prints out " [Lua] "to indicate that the print comes from Lua instead of C++. Next, it loops through each argument and prints them out, separated by a space. Finally, it returns 0 to tell the Lua library that it has no value to return to the call site.

> **To reinforce**
>
> The Lua state and the Lua stack for each `lua_CFunction` call are private to the call. So, everything in the stack is made up of the arguments passed from the Lua call site. You do not need to remove them from the stack before pushing your return values because you are already telling the Lua library how many values are pushed onto the stack as the C++ function return value.

Next, let us see how we can override the Lua `print` function with the C++ version we just implemented.

Overriding the Lua print function

Here, we will first look at the code and then delve into our explanation of it. In the same anonymous namespace, add the following function:

```
namespace
{
    void overrideLuaFunctions(lua_State *L)
    {
        const struct luaL_Reg overrides[] = {
            {"print", luaPrintOverride},
            {NULL, NULL}};
        lua_getglobal(L, "_G");
        luaL_setfuncs(L, overrides, 0);
        lua_pop(L, 1);
    }
}
```

The process of overriding library functions includes the following:

- Getting the library table

- Reassigning the functions of interest to your new implementations

Each line of the code is doing the following:

- It defines an array of `luaL_Reg`, which is a structure representing a name and `lua_CFunction` pair. We set the name as `"print"`, the same as the function name we want to override. We set the function as our new implementation. The last entry in the array must be `{NULL, NULL}` to mark the end of the definition.

- It gets the `_G` Lua table onto the stack, because the `print` function is a global variable and the `_G` table holds all global variables.

- It sets our list of functions from *step 1* to the `_G` table with `luaL_setfuncs`. You can ignore the last parameter for now; we will learn about it in the next section.

- It pops the _G table from the stack to maintain a balanced stack.

- Additionally, `luaL_Reg` is defined in the Lua library as follows:

```
typedef struct luaL_Reg {
    const char *name;
    lua_CFunction func;
} luaL_Reg;
```

Overriding the Lua library functions is really as simple as reassigning some table keys to different values! Now, let us see if it works.

Testing it out

Similar to the previous section, call `overrideLuaFunctions` from the constructor of our Lua executor, as follows:

```
LuaExecutor::LuaExecutor(...)
{
    ...
    overrideLuaFunctions(L);
}
```

You do not need to change anything else. With the same `main.cpp` and `script.lua` files, run the project. If you have followed everything correctly, you should see the following output:

```
[Lua] Host version is 6
```

There is now a `[Lua]` prefix in the output, proving it is printed from our C++ override, not the Lua library function.

Next, let us learn about C++ modules, which is the preferred way to add your C++ functionalities to Lua.

How to register C++ modules

In this section, we will export a C++ class instance to Lua. You probably have used or even implemented Lua modules before, the ones that the Lua interpreter can find and load automatically and return via Lua's `require` function. Here, the focus is integrating Lua into C++, and in such use cases, things are initiated from a C++ executor to benefit from the rest of your C++ application. So, there is a difference if you have used standalone Lua modules before.

In the previous chapter, we implemented a Lua class called `Destinations` to keep track of places we want to go. Let us reimplement it in C++ so that we can export it to Lua.

Implementing a C++ class

Create two source files, `Destinations.h` and `Destinations.cc`. Remember to add `Destinations.cc` to the `Makefile`. Write the header file as follows:

```cpp
#ifndef _DESTINATIONS_H
#define _DESTINATIONS_H

#include <map>
#include <vector>
#include <string>

class Destinations
{
public:
    Destinations(const std::string &name);
    void wish(const std::vector<std::string> &places);
    void went(const std::vector<std::string> &places);
    std::vector<std::string> listVisited() const;
    std::vector<std::string> listUnvisited() const;
private:
    std::string name;
    std::map<std::string, bool> wishlist;
};

#endif // _DESTINATIONS_H
```

We use a map variable to keep a list of places and whether we have visited them and have a name member variable to identify the instance. The member functions are named and work the same as the Lua version, as follows:

- `wish` adds a list of places to the wish list as `unvisited`

- `went` marks a list of places as `visited`

- `listVisited` returns visited places

- `listUnvisited` returns unvisited places

Now, let us implement the member functions in `Destinations.cc`. They are plain C++ functions without using any Lua features. So, we will just list the code without much explanation. First, let us implement the constructor:

```cpp
#include "Destinations.h"

Destinations::Destinations(const std::string &name)
    : name(name), wishlist({}) {}
```

This initializes the wishlist as an empty map.

Then, write the wish function as follows:

```
void Destinations::wish(
    const std::vector<std::string> &places)
{
    for (const auto &place : places)
    {
        wishlist[place] = false;
    }
}
```

Then, implement the went function as follows:

```
void Destinations::went(
    const std::vector<std::string> &places)
{
    for (const auto &place : places)
    {
        wishlist[place] = true;
    }
}
```

The wish function and the went function are quite similar and mark places as visited or unvisited.

Finally, implement the query functions. Write the listVisited function as follows:

```
std::vector<std::string>
Destinations::listVisited() const
{
    std::vector<std::string> results;
    for (const auto &[place, visited] : wishlist)
    {
        if (visited)
        {
            results.push_back(place);
        }
    }
    return results;
}
```

Then, write the listUnvisited function as follows:

```
std::vector<std::string>
Destinations::listUnvisited() const
```

```
{
    std::vector<std::string> results;
    for (const auto &[place, visited] : wishlist)
    {
        if (not visited)
        {
            results.push_back(place);
        }
    }
    return results;
}
```

With a C++ class ready, our next task is to export it to Lua.

What to export to Lua

Exporting a C++ class to Lua is really exporting its instances to Lua. Sometimes, only one instance is exported and the C++ class works as a utility library, similar to the Lua `string` library. Sometimes, many instances are exported and Lua extends C++'s **object-oriented programming (OOP)**.

It is important to note that no matter how many instances you want to export, the process is the same. When overriding Lua library functions, we retrieve an existing table and set some of its functions to our implementation. To export a C++ class instance, similarly, we need to do the following:

- Create a new table
- Add the functions we want to export to the table

If you recall that we can only export functions of the `lua_CFunction` prototype to Lua, you will clearly see that we cannot export our public member functions to Lua directly. We need some wrapper functions. Let us first write some stubs. Below the `#include` directives in `Destinations.cc`, add the following code:

```
namespace
{
int luaWish(lua_State *L) { return 0; }
int luaWent(lua_State *L) { return 0; }
int luaListVisited(lua_State *L) { return 0; }
int luaListUnvisited(lua_State *L) { return 0; }

const std::vector<luaL_Reg> REGS = {
    {"wish", luaWish},
    {"went", luaWent},
    {"list_visited", luaListVisited},
    {"list_unvisited", luaListUnvisited},
```

```
    {NULL, NULL}};
}
```

We defined four wrapper functions of the `lua_CFunction` prototype and a list of `luaL_Reg` instances. We are using `vector` instead of `array` because in C++ we prefer vectors unless we have to use an array.

Next, let us design a reusable mechanism to export our wrappers to Lua.

Designing a reusable exporting mechanism

There are many ways to do this. We choose a way to work with our Lua executor and let it register our C++ modules. First, let us define an abstract class to represent C++ modules. Create a new file named LuaModule.h and write its content as follows:

```
#ifndef _LUA_MODULE_H
#define _LUA_MODULE_H

#include <lua.hpp>
#include <string>
#include <vector>

class LuaModule
{
public:
    virtual const std::string &luaName()
        const override = 0;
    virtual const std::vector<luaL_Reg> &luaRegs()
        const overrode = 0;
    virtual ~LuaModule() = default;
};


#endif // _LUA_MODULE_H
```

The LuaModule abstract class defines two abstract methods to provide data needed to register a C++ module to Lua. luaName returns a name for the module instance; we will use it as the Lua table name. luaRegs returns a list of functions to be exported along with their names in Lua.

Let us extend our Destinations C++ class to conform to this protocol. Change its declaration as follows:

```
class Destinations : public LuaModule
{
public:
    ...
```

```
    const std::string &luaName() const;
    const std::vector<luaL_Reg> &luaRegs() const;
    ...
};
```

After this, add the following implementation to `Destinations.cc`:

```
const std::string &
Destinations::luaName() const
{
    return name;
}

const std::vector<luaL_Reg> &
Destinations::luaRegs() const
{
    return REGS;
}
```

The code simply returns the instance name as `luaName` and the `REGS` we just defined for our stubs as `luaRegs`.

Now comes the time to finally register our C++ class to Lua. In `LuaExecutor.h`, add a function declaration as follows:

```
class LuaExecutor
{
public:
    void registerModule(LuaModule &module);
};
```

The `registerModule` function registers an instance of `LuaModule` to the Lua state that the executor holds.

Next, implement it in `LuaExecutor.cc`:

```
void LuaExecutor::registerModule(LuaModule &module)
{
    lua_createtable(L, 0, module.luaRegs().size() - 1);
    luaL_setfuncs(L, module.luaRegs().data(), 0);
    lua_setglobal(L, module.luaName().c_str());
}
```

This needs some explanation. Let us explore what each line of the code here is doing in sequence:

1. It creates a table with `lua_createtable`. This library function will push the table onto the stack. The second parameter hints at how many elements in the table will be used as a sequence. We have none, so we pass 0. The third parameter hints at how many elements in the table will be used as a map. All our functions are used this way, so we pass the count of our vector minus the ending marker. The hints help with memory allocation in Lua, as Lua will be responsible for creating a properly sized table to avoid unnecessary reallocations to increase the capacity of the table later.

2. It sets our functions to the table with `luaL_setfuncs`. This works exactly the same as when we overrode Lua library functions. Ignore the third parameter for now as well. `module.luaRegs().data()` returns our function list as an array instead of a vector. `std::vector::data` is a C++ feature.

3. It assigns the table just created to a global variable using the name returned from `module.luaName()`. From now on, our C++ module can be accessed from this table.

Exporting a C++ module to Lua may sound heavy and glorious. But there is actually not much gluing code involved. Compare what we have just done with overriding Lua library functions. Take a moment, then we will test our mechanism to see if it works.

Testing our mechanism

Add a few lines of code to `main.cpp` so that it looks like this:

```cpp
#include "LuaExecutor.h"
#include "LoggingLuaExecutorListener.h"
#include "Destinations.h"

int main()
{
    auto listener = std::make_unique<
        LoggingLuaExecutorListener>();
    auto lua = std::make_unique<LuaExecutor>(*listener);
    auto wishlist = std::make_unique<Destinations>(
        "destinations");
    lua->registerModule(*wishlist.get());
    lua->executeFile("script.lua");
    return 0;
}
```

We create an instance of the `Destinations` class and give it the name `"destinations"`. Then, we register it with our Lua executor.

Now, add the following code to `script.lua`:

```
destinations.wish("London", "Paris", "Amsterdam")
destinations.went("Paris")
print("Visited:", destinations.list_visited())
print("Unvisited:", destinations.list_unvisited())
```

This is doing the following:

1. It adds London, Paris, and Amsterdam to the wish list.
2. It marks Paris as visited.
3. It prints the visited cities.
4. It prints the unvisited cities.

Run the project, and if you have followed all the steps correctly, you should see the following output:

```
[Lua] Visited:
[Lua] Unvisited:
```

There should be no errors, and since our wrapper functions are only stubs, it will not return anything useful. Hence, this is how we laid out the architectural foundation. Next, we will focus our efforts on making it work at the ground level.

Accessing the C++ class instance

Our wrapper functions are of type `lua_CFunction`. They are essentially C++ static methods not associated with any class. How can we access the correct class instance? We must do some bookkeeping.

Luckily, Lua provides a mechanism to keep data for the registered C++ functions. It is called an **upvalue**. Upvalues can only be accessed by the associated function in C/C++ code and are shared across different function calls. We can save the pointer to the class instance in an upvalue.

Why is it called an upvalue? At this stage, it is easier to understand when not explained, in the same spirit as why a variable is called a variable.

> **Have you noticed?**
>
> From the previous description, an upvalue behaves like a C++ static variable in the function scope. Then, why do we use an upvalue instead of a static variable? Because an upvalue is associated with a C++ function in the Lua library. This way, we can use the same C++ function with different upvalues.

Which Lua data type can be used to save a C++ pointer? We can use **userdata**. This type is used to store arbitrary C/C++ data. Especially, for our case, we need to use **light userdata**, whose purpose is to store a C/C++ pointer. It is a perfect match for us.

In conclusion, we need to save the class instance's this pointer as *light userdata* in an *upvalue* for the lua_CFunction implementation.

Here, we have involved two new Lua concepts. They are exclusively used to work with C/C++ code, so chances are that you are not very familiar with them from Lua programming. Let us see the code in action to help with the understanding.

How to provide upvalues

We will only look at the case for registering C++ modules. So far, we have ignored the third parameter to luaL_setfuncs and always passed 0.

What does this third parameter mean? It is the count for the upvalues that will be available to all functions in the list provided in the second parameter.

How do you provide upvalues? Of course—you push them onto the stack!

Let us rewrite the function to register C++ modules as follows:

```
void LuaExecutor::registerModule(LuaModule &module)
{
    lua_createtable(L, 0, module.luaRegs().size() - 1);
    int nUpvalues = module.pushLuaUpvalues(L);
    luaL_setfuncs(L, module.luaRegs().data(), nUpvalues);
    lua_setglobal(L, module.luaName().c_str());
}
```

There are only two changes. First, we want another function yet to be implemented in LuaModule to push upvalues onto the stack and return to us how many upvalues have been pushed. Then, we pass the upvalue count as the third parameter to luaL_setfuncs.

Remember to add pushLuaUpvalues to LuaModule.h, like so:

```
class LuaModule
{
public:
    virtual int pushLuaUpvalues(lua_State *L)
    {
        lua_pushlightuserdata(L, this);
        return 1;
    }
};
```

We have provided a default implementation that pushes this as an upvalue. In derived classes, they can override this function and push more upvalues.

Next, let us see how we can access this upvalue.

How to access upvalues

Lua upvalues are accessed as if they were in the stack, while they are not really in the stack. So, a magic stack index, LUA_REGISTRYINDEX, is used to mark the start of the upvalue pseudo-region. Lua provides a lua_upvalueindex macro to locate the indices of your upvalues, so you do not really need to deal with this magic number.

This is how we can access our C++ class instance stored as an upvalue. In Destinations.cc, add the following function to the anonymous namespace:

```
namespace
{
    inline Destinations *getObj(lua_State *L)
    {
        return reinterpret_cast<Destinations *>(
            lua_touserdata(L, lua_upvalueindex(1)));
    }
}
```

We can use this helper function to get a pointer to the instance. It uses lua_touserdata to get our light userdata from the stack with the pseudo-index. This helper will be called from the stubs we registered.

> **To reinforce**
>
> The Lua state and stack passed to a lua_CFunction function are private to each call to that function.

Now that we have figured out how to access the class instance, we can complete our stubs.

Completing our stubs

Write luaWish as follows:

```
int luaWish(lua_State *L)
{
    Destinations *obj = getObj(L);
    std::vector<std::string> places;
    int nArgs = lua_gettop(L);
    for (int i = 1; i <= nArgs; i++)
    {
        places.push_back(lua_tostring(L, i));
    }
```

```
    obj->wish(places);
    return 0;
}
```

It first gets the class instance with the `getObj` helper function we just implemented. Then, it puts all arguments from the Lua call site into a vector. Finally, it calls the real object method, `obj->wish`. This is what a wrapper does – it routes the call to the real object.

The code for `luaWent` is similar, as we can see here:

```
int luaWent(lua_State *L)
{
    Destinations *obj = getObj(L);
    std::vector<std::string> places;
    int nArgs = lua_gettop(L);
    for (int i = 1; i <= nArgs; i++)
    {
        places.push_back(lua_tostring(L, i));
    }
    obj->went(places);
    return 0;
}
```

The only difference is that it calls `obj->went` instead.

Finally, implement the query functions as follows:

```
int luaListVisited(lua_State *L)
{
    Destinations *obj = getObj(L);
    auto places = obj->listVisited();
    for (const auto &place : places)
    {
        lua_pushstring(L, place.c_str());
    }
    return places.size();
}

int luaListUnvisited(lua_State *L)
{
    Destinations *obj = getObj(L);
    auto places = obj->listUnvisited();
    for (const auto &place : places)
    {
        lua_pushstring(L, place.c_str());
```

```
    }
    return places.size();
}
```

These functions use the object functions to get a list of places and then push the list onto the stack to return the results to the Lua call site.

Now, we have implemented everything, and we can test it.

Testing it out

We do not need to modify any test code because we have already used the functions to test our stubs. Now, recompile and run the project.

Recall that the Lua test code looks like this:

```
destinations.wish("London", "Paris", "Amsterdam")
destinations.went("Paris")
print("Visited:", destinations.list_visited())
print("Unvisited:", destinations.list_unvisited())
```

If you have done everything correctly, you should see the following output:

```
[Lua] Visited: Paris
[Lua] Unvisited: Amsterdam London
```

Congratulations on making it work!

This chapter is quite a change of mindset from the previous chapters. Take a moment to reflect if you need to.

Summary

In this chapter, we learned how to call C++ code from Lua. We first learned how to register a simple C++ function to Lua. All registered functions must conform to `lua_CFunction`. Then, we found out how to override Lua library functions. Finally, we implemented a C++ class and exported it to Lua. We also came across the concepts of *upvalue* and *light userdata* along the way.

In the next chapter, we will continue our journey with more details on user-defined data in C++ and more data-exchanging mechanisms.

Exercises

1. In the Destinations class, we only used one upvalue. Add another upvalue and play around with it. Which upvalue is at which pseudo-index?

2. Try to modify the second upvalue in a function and see if the value is persisted the next time the function is called. How about when it is accessed in another function?

3. In LuaType.hpp, add LuaType::lightuserdata and implement a structure for it, named LuaLightUserData. Support this case in the executor and helper functions. You do not need to support this type when popping values from the Lua stack.

7
Working with C++ Types

In the previous chapter, we learned how to call C++ code from Lua and how to register C++ modules. You might have observed that you can create objects in C++ and register them to Lua, but how about creating C++ objects freely in Lua? This is what we will set out to achieve in this chapter.

We will learn about the following topics and export C++ types to Lua in a top-down approach:

- How to use the Lua registry
- How to use userdata
- Exporting C++ type to Lua

Technical requirements

Here are the technical requirements for this chapter:

- You can access the source code for this chapter at `https://github.com/ PacktPublishing/Integrate-Lua-with-CPP/tree/main/Chapter07`.
- The previous chapter showed you how to register C++ modules; you must have finished coding the questions. You can check out the answers in the `begin` folder, which can be found in this book's GitHub repository.
- In this chapter, due to its complexity, we are adopting a top-down approach so that you only get a working implementation toward the end of this chapter. You can always check the end folder in this book's GitHub repository if you prefer to see the complete code from the beginning.

How to use the Lua registry

The **Lua registry** is a global table that can be accessed by all C/C++ code. It is one of the places that C++ code can keep state across different function calls. You cannot access this table in Lua code unless you use the Lua *debug* library. However, you should not use the debug library in production code; so, the registry is for C/C++ code only.

> **The registry versus upvalues**
>
> In the previous chapter, we learned about Lua *upvalues*. An *upvalue* keeps a state for a specific Lua C function across calls. The *registry*, on the other hand, can be accessed by all Lua C functions.

To export C++ types to Lua, we will use Lua userdata to represent the type and the registry so that we have a metatable for functions for the type. We will learn about the registry first, then the userdata, and finally put everything together to export C++ types to Lua.

Let's add support for the registry in our Lua executor so that we know how to use it.

Supporting the registry

Since the registry is a table, we must get a value with a key and set a value with a key. We can use `LuaValue` to represent different types of Lua values.

In `LuaExecutor.h`, add the following function declarations:

```
class LuaExecutor
{
public:
    LuaValue getRegistry(const LuaValue &key);
    void setRegistry(const LuaValue &key,
                     const LuaValue &value);
};
```

`getRegistry` will return the value from the registry for `key`. `setRegistry` will set `value` with `key` to the registry.

Now, let's implement them and learn which Lua library functions can be used. In `LuaExecutor.cc`, implement `getRegistry`, as follows:

```
LuaValue LuaExecutor::getRegistry(const LuaValue &key)
{
    pushValue(key);
    lua_gettable(L, LUA_REGISTRYINDEX);
    return popValue();
}
```

This looks simple, right? It reuses two pieces of knowledge we have come across from the previous chapters:

- We use `lua_gettable` to get a value from a table, where the top of the stack is the key and the table is located in the position in the stack specified by the function parameter. We learned about this in *Chapter 5*.

- Similar to upvalues, the registry is a special use case regarding the Lua stack, so it also has a pseudo-index called LUA_REGISTRYINDEX. We first encountered this pseudo-index in *Chapter 6*.

- By combining these two points, we get an even simpler implementation compared to getting a value for a normal table. This is because we do not need to push the table onto the stack.

Next, we will implement setRegistry. In LuaExecutor.cc, add the following code:

```
void LuaExecutor::setRegistry(const LuaValue &key,
                              const LuaValue &value)
{
    pushValue(key);
    pushValue(value);
    lua_settable(L, LUA_REGISTRYINDEX);
}
```

This only needs to call one Lua library function: lua_settable. value is located on top of the stack and key is located second to the top. This simplicity is due to the great design of the Lua API regarding the pseudo-index.

With our Lua executor extended, let's test it to see how the registry works.

Testing the registry

In main.cpp, replace the test code, as follows:

```
int main()
{
    auto listener = std::make_unique<
        LoggingLuaExecutorListener>();
    auto lua = std::make_unique<LuaExecutor>(*listener);

    auto key = LuaString::make("darabumba");
    lua->setRegistry(key,
        LuaString::make("gwentuklutar"));
    auto v1 = lua->getRegistry(key);

    lua->setRegistry(key, LuaString::make("wanghaoran"));
    auto v2 = lua->getRegistry(key);

    std::cout << getLuaValueString(key)
              << " -> " << getLuaValueString(v1)
              << " -> " << getLuaValueString(v2);
    return 0;
}
```

The test code is doing the following, separated by newlines:

- Creates LuaExecutor and a listener, as usual
- Sets the registry with a key-value pair using some strings
- Sets the key to another value
- Prints out the key, the initial value, and the current value

If you have done everything correctly, you should see the following output:

```
darabumba -> gwentuklutar -> wanghaoran
```

In the previous chapter, we used an upvalue to store a light userdata. Now, let's test light userdata with the registry. Replace the registry operations with the following:

```
auto regkey = LuaLightUserData::make(listener.get());
lua->setRegistry(regkey, LuaString::make(
    "a LuaExecutorListener implementation"));
auto regValue = lua->getRegistry(regkey);
std::cout << std::hex << listener.get() << " is "
          << getLuaValueString(regValue);
```

Here, we used light userdata as the key and a string as a value explaining what the key is. You should see an output similar to the following:

```
0x14f7040d0 is a LuaExecutorListener implementation
```

Your key address will differ in each run.

Now that we've covered the registry and reviewed light userdata, let's learn about Lua userdata.

How to use userdata

From this section onwards, we will transform the Destinations class from the previous chapter and register it to Lua as a type so that Lua code can create objects from it. Before we dive into the details, let's make some high-level changes to show what we want at the project level.

Preparing the C++ type

In the previous chapter, we passed a name to the constructor for the Destinations class because we created its instances in C++ and needed to set a name to Lua for each instance.

In this chapter, we will export the C++ type to Lua. Lua will create the object and assign it a name. All changes will be in the Destinations class. The mechanism we implemented in the Lua executor to register C++ modules is flexible enough to support the registration of C++ types as well.

To reflect this change and difference, we will change the constructor and how the module name is provided. In `Destinations.h`, change the constructor, as follows:

```
class Destinations : public LuaModule
{
public:
    Destinations();
};
```

Then, delete the following private member variable:

```
class Destinations : public LuaModule
private:
    std::string name;
};
```

We will register a type instead of an instance. We can use a static variable for the Lua table name. You can change the constructor in `Destinations.cc` accordingly:

```
Destinations::Destinations() : wishlist({}) {}
```

Now, let's reimplement how to provide the Lua type/table name. In `Destinations.cc`, change `luaName` so that it uses a static variable:

```
namespace
{
    const std::string NAME("Destinations");
}
const std::string &Destinations::luaName() const
{
    return NAME;
}
```

Add a string constant called NAME with a value of `Destinations` at the beginning of the anonymous namespace and return it in `luaName`.

Finally, change the test code in `main.cpp`, as follows:

```
int main()
{
    auto listener = std::make_unique<
        LoggingLuaExecutorListener>();
    auto lua = std::make_unique<LuaExecutor>(*listener);
    auto wishlist = std::make_unique<Destinations>();
    lua->registerModule(*wishlist.get());
    lua->executeFile("script.lua");
```

```
    return 0;
}
```

Compared to the code at the end of *Chapter 6*, the only change here is that we removed the parameter to the constructor for the `Destinations` class.

This is what we want on the C++ side. Make sure it compiles. From now on, we will focus on re-wrapping the `Destinations` class and making `script.lua` create objects from it.

Next, let's learn about userdata.

What is userdata?

In the previous chapter, we exported an instance of the class as a table and the class member functions as table functions. To export a type directly, we will create a userdata instead of a table.

In *Chapter 2*, we learned that userdata is one of the basic types in Lua, but we did not dive into the details. In the previous chapter, we used lightuserdata. Is there any difference between userdata and lightuserdata?

A userdata is an arbitrary sequence of data that's created from C/C++ code by calling the Lua library and Lua treats it transparently. On the other hand, lightuserdata must be a C/C++ pointer.

What makes userdata more suitable to represent new types, is that, like a table, you can set metamethods and metatables to it. This is how you provide type functions and make it a useful type in Lua.

> **Object-oriented programming in Lua**
>
> In *Chapter 5*, we learned about object-oriented programming in Lua and the __index metatable. If this sounds unfamiliar, please revisit that chapter before continuing.

Now, let's look at what we should put into the userdata so that we can export C++ types.

Designing the userdata

To create userdata, Lua provides the following function:

```
void *lua_newuserdatauv (
    lua_State *L, size_t size, int nuvalue);
```

This function allocates a consecutive memory block with a length of `size` bytes as a userdata and pushes a reference to the userdata onto the stack. You can also attach user values to the userdata. The count of user values is passed in `nuvalue`. We will not use user values, so we will pass 0. `lua_newuserdatauv` returns the address of the raw memory that's been allocated.

Since Lua is written in C, you can use `lua_newuserdatauv` to allocate a C array or structure. Lua will even take care of deallocating it during garbage collection.

With C++, we want the userdata to represent a class. It is not portable or convenient to call this Lua library function to allocate a C++ class. So, we will take matters into our own hands – we will create the C++ object on our own and make the userdata a pointer to the object.

> **Decoupling**
>
> Although we are integrating Lua into C++, we have chosen to decouple the C++ side and the Lua side as much as possible and only expose the necessary interfaces. C++ memory management is a complex topic already. We have chosen to let C++ manage the C++ object creation and only use Lua userdata to keep a pointer and as a garbage collection signal.

Let's start by writing a Lua C function for object creation. In `Destinations.cc`, at the end of the anonymous namespace, add a function called `luaNew`. Add it to the `REGS` vector as well:

```
int luaNew(lua_State *L);
const std::vector<luaL_Reg> REGS = {
    {"new", luaNew},
    ...
    {NULL, NULL}};

int luaNew(lua_State *L)
{
    Destinations *obj = new Destinations();
    Destinations **userdata =
        reinterpret_cast<Destinations **>(
            lua_newuserdatauv(L, sizeof(obj), 0));
    *userdata = obj;
    return 1;
}
```

`luaNew` will be responsible for creating `Destinations` instances. This is a three-step process:

First, we create an instance of the class in the heap with new and store its pointer in `obj`.

Then, we call `lua_newuserdatauv` to create a userdata to hold the pointer in `obj`. `userdata` will have a size of `sizeof(obj)`, which is the size of a C++ object pointer. Because `lua_newuserdatauv` returns the pointer to the raw memory and we have made this memory hold a pointer to the `Destinations` instance, we need to save the address of the allocated memory as a pointer to a pointer.

Finally, we make `*userdata` point to `obj`. Since `userdata` is already on top of the stack, we can return 1 to return the allocated userdata to Lua.

When Lua code creates a Destinations instance via luaNew, the Lua side will get a userdata.

A pointer to a pointer

C++ allows you to create a pointer to a pointer. In such a case, one pointer holds the address of the other pointer. This is not frequently used in pure C++ code but can be useful for interacting with C APIs. The Lua library is a C library.

Next, let's prepare script.lua so that it can use the C++ type.

Preparing the Lua script

Replace the content of script.lua, as follows:

```lua
dst = Destinations.new()
dst:wish("London", "Paris", "Amsterdam")
dst:went("Paris")
print("Visited:", dst:list_visited())
print("Unvisited:", dst:list_unvisited())

dst = Destinations.new()
dst:wish("Beijing")
dst:went("Berlin")
print("Visited:", dst:list_visited())
print("Unvisited:", dst:list_unvisited())
```

This first part of the script is similar to the script we used for the previous chapter, except that we changed to Destinations.new() with the new table name and we switched to use the object calling convention with colons. The second part of the script is a repetition of the first part with different city names. This is to demonstrate that we can create many Destinations instances in Lua.

If you run the project at this point, you will see the following error:

```
[LuaExecutor] Failed to execute: script.lua:2: attempt to index a
userdata value (global 'dst')
```

For now, this is expected. Because we returned a userdata instead of a table to Lua, so far, it is just a raw piece of memory that's transparent to Lua. Lua does not know how to call a method on a piece of raw memory. As explained earlier, we need to set a metatable to the userdata for this to work. We will do this next to put everything together.

Exporting C++ types to Lua

In the previous section, we returned the pointer to the class instance to Lua as a userdata. In this section, we will export member functions for the userdata. To do this, we need to set the metatable for the userdata.

Setting a metatable for the userdata

In *Chapter 5*, we learned that each table can have a metatable. Similarly, each userdata can also have a metatable. In *Chapter 2*, we learned that in Lua code, the metatable needs to be set during object creation.

Here, we need to set the metatable during object creation in C++ code. Instead of creating a new metatable for each object, we can create a single metatable and store it in the registry. Then, each object only needs to reference this single metatable. This will increase efficiency and reduce memory footprint. The Lua library even provides helper functions for this.

First, let's see the code; an explanation will follow. Replace the content for luaNew, as follows:

```
const std::string METATABLE_NAME(
    "Destinations.Metatable");

int luaNew(lua_State *L)
{
    Destinations *obj = new Destinations();
    Destinations **userdata =
        reinterpret_cast<Destinations **>(
            lua_newuserdatauv(L, sizeof(obj), 0));
    *userdata = obj;

    int type = luaL_getmetatable(
        L, METATABLE_NAME.c_str());
    if (type == LUA_TNIL)
    {
        lua_pop(L, 1);
        luaL_newmetatable(L, METATABLE_NAME.c_str());
        lua_pushvalue(L, -1);
        lua_setfield(L, -2, "__index");
        luaL_setfuncs(L, REGS.data(), 0);
    }
    lua_setmetatable(L, 1);

    return 1;
}
```

As you can see, the code here is separated by newlines. In this case, we have only inserted the second section. The rest of the code is kept as-is. Additionally, we declared a new constant called METATABLE_NAME. You can put this after the NAME constant. The second code section does the following:

1. First, it gets the metatable containing the METATABLE_NAME key from the registry. The luaL_getmetatable library function is used for this.

2. If it's not found, the metatable is created. We will expand on this detail later.

3. Finally, the code section sets the metatable to the userdata with lua_setmetatable. This library function expects the metatable to be on top of the stack. In our case, the position of the userdata in the stack is specified via the 1 parameter. lua_setmetatable will pop the metatable from the stack.

To understand this better, see the following figure:

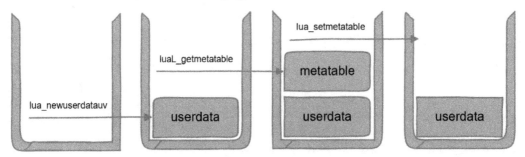

Figure 7.1 – Setting the metatable

The preceding figure shows how the stack's state changes when the metatable is already in the registry.

Next, let's look at the case when the metatable is not in the registry.

Creating a metatable for the userdata

In our implementation, the metatable is created when a class instance is created from Lua for the first time. Specifically, the code to create the metatable is as follows; this was copied from the previous code listing:

```
if (type == LUA_TNIL)
{
    lua_pop(L, 1);
    luaL_newmetatable(L, METATABLE_NAME.c_str());
    lua_pushvalue(L, -1);
    lua_setfield(L, -2, "__index");
    luaL_setfuncs(L, REGS.data(), 0);
}
```

When we need to create the metatable, the stack's state changes, as shown in the following figure:

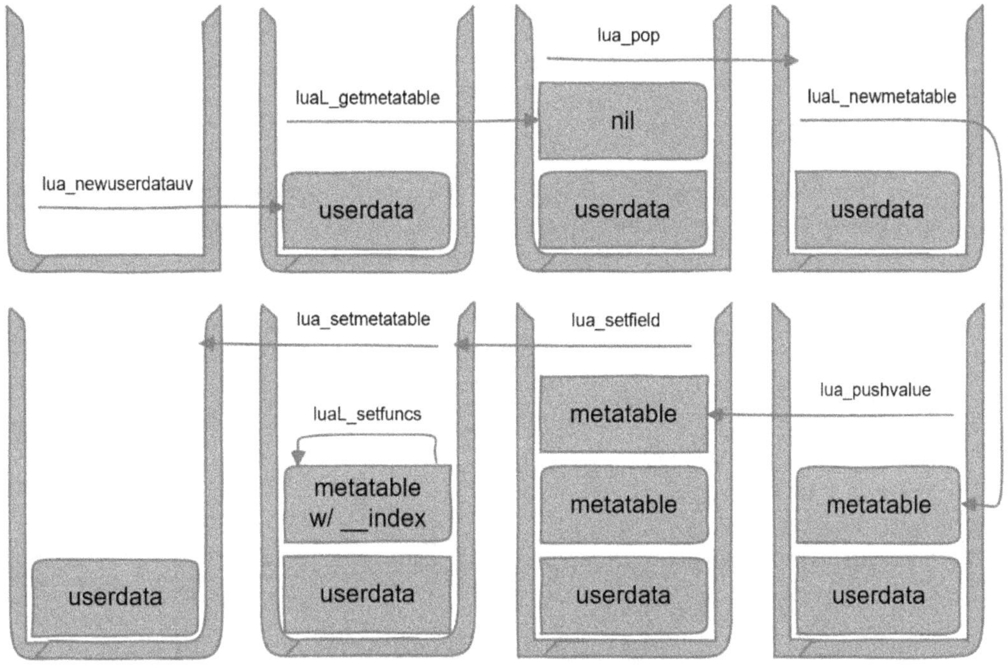

Figure 7.2 – Creating the metatable

Eight steps are involved in this process. Steps 4 to 7 deal with creating the new metatable.

1. Creates a userdata with `lua_newuserdatauv` that holds a pointer to the class instance.

2. Tries to get a metatable from the registry with `luaL_getmetatable`. The metatable does not exist yet, so we will get a nil value.

3. Pops the nil value from the stack with `lua_pop`. Because it is not useful and is on top of the userdata in the stack, we need to return the userdata as the top of the stack.

4. Creates an empty metatable in the registry with `luaL_newmetatable`.

5. Pushes a copy of the reference to the metatable onto the stack with `lua pushvalue`. Now, we have two references to the same metatable; we will use these in the next step.

6. Sets the `__index` field of the metatable as itself with `lua_setfield`. This ensures that we do not need to create another table for the `__index` field.

7. Sets the type functions in REGS to the metatable with `luaL_setfuncs`.

8. Sets the metatable to the userdata with `lua_setmetatable`.

With this, when the Lua code calls a method on the userdata object, it will dispatch to the proper Lua C function wrapper.

How can we get the object in C++ in the Lua C wrapper function?

In the previous chapter, we used upvalue. Although this can still work, it is no longer suitable here. Let's make object retrieval in C++ work again.

Getting the object in C++

In `script.lua`, we have changed the Lua code so that it uses the object calling convention with colons. This means that the object will be passed to the Lua C functions as the first argument.

To get the object reference, rewrite the `getObj` function, as follows:

```
inline Destinations *getObj(lua_State *L)
{
    luaL_checkudata(L, 1, METATABLE_NAME.c_str());
    return *reinterpret_cast<Destinations **>(
        lua_touserdata(L, 1));
}
```

This function is called from Lua C wrapper functions to get the object reference. This was first implemented in the previous chapter using an upvalue.

First, using the Lua `luaL_checkudata` library function, we check that the first argument is a userdata with a metatable in the `METATABLE_NAME` registry. This is a security measure to ensure that the Lua code does not pass an object of another type. The check on the metatable's name works because only C/C++ code can access the registry.

Then, we return the first argument as a userdata cast to `Destinations **` and dereferenced to get the object pointer.

Since we are now passing the object from Lua in the calls, we need to modify our wrapper functions slightly.

Making wrappers work again

In `luaWish` and `luaWent`, we were expecting a list of cities. Now, we need to exclude the first argument. There is only one character to change. Rewrite `luaWish`, as follows:

```
int luaWish(lua_State *L)
{
    Destinations *obj = getObj(L);
    std::vector<std::string> places;
    int nArgs = lua_gettop(L);
    for (int i = 2; i <= nArgs; i++)
    {
        places.push_back(lua_tostring(L, i));
    }
```

```
    obj->wish(places);
    return 0;
}
```

We only changed `i = 1` to `i = 2` for the `for` loop. You can do the same for `luaWent`.

Now that we have exported the `Destinations` C++ type to Lua, let's test it.

Testing it out

Since we have adopted a top-down approach in this chapter, all the test code is complete. Now, we only need to run the project.

If you've been following along, you should see the following output:

```
[Lua] Visited: Paris
[Lua] Unvisited: Amsterdam London
[Lua] Visited: Berlin
[Lua] Unvisited: Beijing
```

The first two lines are from the first object that was created in `script.lua`. The last two lines are from the second object that was created.

If you have made any errors in the code, you will see some error output here. Go back and see what you have missed. In this chapter, the code is a continuation of the previous chapter, but the stack's state is quite complex.

Should we say, "Congratulations"? Wait a moment; have you discovered something we missed in the exported type?

Let's print something in the constructor and the destructor to trace the object's life cycle. Rewrite the constructor and the destructor, as follows:

```
Destinations::Destinations() : wishlist({})
{
    std::cout << "Destinations instance created: "
              << std::hex << this << std::endl;
}

Destinations::~Destinations()
{
    std::cout << "Destinations instance destroyed: "
              << std::hex << this << std::endl;
}
```

The constructor and the destructor simply output the object pointer value in hex format.

Run the project again and see what you get. You should see something similar to the following output:

```
Destinations instance created: 0x12df07150
Destinations instance created: 0x12f804170
[Lua] Visited: Paris
[Lua] Unvisited: Amsterdam London
Destinations instance created: 0x12f8043a0
[Lua] Visited: Berlin
[Lua] Unvisited: Beijing
Destinations instance destroyed: 0x12df07150
```

Three instances are created, but only one is destroyed! The one that gets destroyed is the one that was created in main.cpp and was serving as the prototype for the Destinations type. This means that we are leaking objects in Lua!

We need to fix this!

Providing a finalizer

Are you wondering why there is a memory leak?

This is because Lua uses garbage collection. We allocate a userdata and use it as a pointer. It is this pointer that gets garbage collected when the userdata is out of scope in Lua. The C++ object itself is not deleted.

Lua supports using userdata in this way perfectly fine, but you need to help Lua delete the object when the associated userdata is garbage collected. To help Lua, you can provide a __gc metamethod in the metatable. This is called a **finalizer**. It is invoked during the garbage collection process to delete your real object.

Let's provide a finalizer named luaDelete for the Destinations type. Add another Lua C function above luaNew:

```
int luaDelete(lua_State *L)
{
    Destinations *obj = getObj(L);
    delete obj;
    return 0;
}
```

Easy, right? Lua passes the userdata object to the finalizer during garbage collection.

Now, let's register it. In luaNew, add two more lines, as follows:

```
int luaNew(lua_State *L)
{
    ....
```

```
    if (type == LUA_TNIL)
    {
        ...
        lua_pushcfunction(L, luaDelete);
        lua_setfield(L, -2, "__gc");
    }
    ...
}
```

This sets the __gc metamethod for the metatable to our luaDelete function.

The interactions between Lua and C++ can be seen in the following diagram:

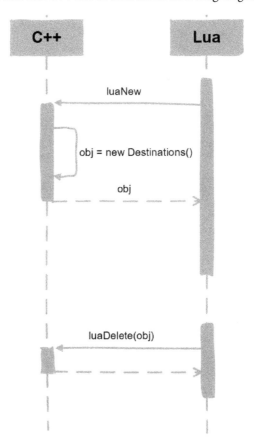

Figure 7.3 – Object creation and destruction

When Lua code creates an object of the Destinations class, it triggers a call to luaNew, which creates the C++ object and returns its pointer as a userdata. When Lua is done with the object, after a while, garbage collection is triggered and the luaDelete C++ finalizer is called.

Let's run the project again. You should see an output similar to the following:

```
Destinations instance created: 0x14c609a60
Destinations instance created: 0x14c704170
[Lua] Visited: Paris
[Lua] Unvisited: Amsterdam London
Destinations instance created: 0x14c7043a0
[Lua] Visited: Berlin
[Lua] Unvisited: Beijing
Destinations instance destroyed: 0x14c609a60
Destinations instance destroyed: 0x14c7043a0
Destinations instance destroyed: 0x14c704170
```

With that, we have created three instances and destroyed three instances.

Congratulations!

Summary

In this chapter, we learned how to export C++ types to Lua so that Lua code can create C++ objects. We also learned about the registry and the userdata type. Last but not least, we implemented a finalizer for garbage collection.

The work in this chapter is based on registering C++ modules from the previous chapter. You can choose to export a C++ class or C++ class instances. It is a matter of design choice and your project's needs.

In the next chapter, we will learn how to implement a template class to help export C++ types to Lua.

Exercise

In main.cpp, we created an instance of Destinations and registered it to Lua. In Lua code, this instance is used to create more instances. This is fine, considering that Lua uses prototype-based object-oriented programming as well. If you want to make your C++ code more like C++'s way of doing things, you can create a factory class for Destinations and register the factory to Lua. Do this and make the factory only export the luaNew function. You shouldn't need to change script. lua or the Lua C wrapper function implementations.

8

Abstracting a C++ Type Exporter

In the previous chapter, we learned how to export a C++ class to Lua as a user-defined type. An instance of the class is exported to Lua to serve as the prototype for the type.

The exercise from the previous chapter required you to create a factory class for the type. However, with that requirement, each type needs its own factory class.

In this chapter, we will learn how to implement a general C++ type exporter so that you can use it as the factory class for any C++ type, without redoing the work.

In this chapter, we will cover the following topics:

- Reviewing the factory implementation
- Designing a type exporter
- Stubbing the type exporter
- Defining LuaModuleDef
- Re-implementing luaNew
- Are you flexible enough?

Technical requirements

You can access the source code for this chapter here: https://github.com/PacktPublishing/Integrate-Lua-with-CPP/tree/main/Chapter08.

Reviewing the factory implementation

We will review the exercise from the previous chapter. The solution this book has adopted requires progressive and minimal changes. It also leads naturally to the feature covered in this chapter.

If you have not implemented your own solution and are willing to stop here for a moment, this is another chance to give it a shot. Many technologies are simple in explanation but hard to grasp. The best way to understand them is to review and practice them, hands-on, again and again until you get an "aha" moment.

Now, we will review a solution. The focus is on the changes and key concepts.

Defining the factory

To make a factory, we only need to change `Destinations.h` and `Destinations.cc`. In your preferred IDE, you can open the end project for *Chapter 7* and the `begin` project for this chapter to check the differences.

Let us first look at the header file for the factory class declaration. You can find the following declarations in `Destinations.h` in the `begin` project for this chapter:

```
class Destinations
{
public:
    Destinations();
    ~Destinations();
    void wish(const std::vector<std::string> &places);
    void went(const std::vector<std::string> &places);
    std::vector<std::string> listVisited() const;
    std::vector<std::string> listUnvisited() const;
private:
    std::map<std::string, bool> wishlist;
};

class DestinationsFactory : public LuaModule
{
public:
    const std::string &luaName() const override;
    const std::vector<luaL_Reg> &
    luaRegs() const override;
};
```

The change from the previous chapter is that we are creating a factory class named `DestinationsFactory`, which implements the `LuaModule` interface. In fact, we are moving the `LuaModule` implementation from `Destinations` to `DestinationsFactory` so that the `Destinations` type does not know anything about Lua. This is one of the benefits of a factory class. The system can be better layered.

Did you know?

If you use Linux or Mac, you can also use the **diff** program to check the differences between two files. In a terminal, type the following command to see what we have changed – `diff Chapter07/end/Destinations.h Chapter08/begin/Destinations.h`.

Next, we will review the factory implementation.

Implementing the factory

The factory only has two member functions, which are implemented as follows:

```
const std::string &
DestinationsFactory::luaName() const
{
    return NAME;
}

const std::vector<luaL_Reg> &
DestinationsFactory::luaRegs() const
{
    return FACTORY_REGS;
}
```

Now, instead of REGS, luaRegs returns FACTORY_REGS, which is defined as follows:

```
const std::vector<luaL_Reg> FACTORY_REGS = {
    {"new", luaNew},
    {NULL, NULL}};
```

This means, now, we only export a single function, luaNew, to Lua.

As explained in *Chapter 6*, the Lua library expects the last entry to be {NULL, NULL} to mark the end of the array. This is a typical technique for C-based libraries because they usually get a pointer to an item as the input for an array and need to figure out where the array ends.

Also, delete luaNew from REGS so that it looks like the following list:

```
const std::vector<luaL_Reg> REGS = {
    {"wish", luaWish},
    {"went", luaWent},
    {"list_visited", luaListVisited},
    {"list_unvisited", luaListUnvisited},
    {NULL, NULL}};
```

Previously, `REGS` served two purposes:

- The `__index` metatable for new instances created from Lua. This is done in `luaNew` with `luaL_setfuncs(L, REGS.data(), 0)`.
- The registered Lua module, which is a plain Lua table. This is done with a call to `LuaExecutor::registerModule`.

Now, `REGS` only serves the first purpose and gives the second responsibility to `FACTORY_REGS`. This is yet another structural improvement.

These are all the changes we need to create a factory. You can read the complete source code that you get from GitHub. However, there is not much code change, right? We were just moving things around, and we now have a different object-creation mechanism.

Now, based on this factory concept, we are ready to move on to the main focus of this chapter. From now on, you can use the `begin` project as the basis for development. Let us start to design a general C++ type exporter.

Designing a type exporter

First, let us define our scope. We want to generalize the factory we have just made and make it work with any C++ class – that is, the C++ class still needs to implement and provide the `lua_CFunction` wrappers in some way. It is possible to automate the creation of those wrappers, but that would require implementing a heavy C++ template library, which is not strictly related to Lua and is out of the scope of this book.

With the scope defined, let us make some high-level designs.

Choosing a design pattern

When we talk about making something *general* in C++, usually it means we need to use templates. To work with our Lua executor, we need to export `LuaModule`. So, we need to implement the exporter as a template class that can provide `LuaModule`.

How can we provide `LuaModule`? We can make the exporter inherit from the `LuaModule` interface, or make one of its member functions return `LuaModule`.

One of the popular design patterns for the latter option is the **Builder** pattern. This can be demonstrated with the following pseudo-code:

```
class LuaModuleBuilder
{
    LuaModuleBuilder withOptionA(...);
    LuaModuleBuilder withOptionB(...);
    ...
    LuaModule build();
```

```
};

LuaModuleBuilder builder;
auto module = builder.withOptionX(...).build();
```

A builder usually has many functions to customize different properties for the thing it creates, alongside a `build` function to create the final object.

Since our goal is to help with object creation alone, like in the factory exercise, and not to customize the object, the *Builder* pattern is overkill. We will choose the vanilla C++ inheritance. The exporter type can be defined as follows:

```
template <typename T>
class LuaModuleExporter : public LuaModule;
```

This is a template class. It will export C++ type T as `LuaModule`.

Now, let us stub the exporter.

Stubbing the exporter

We have two major considerations for the design of the exporter. First, it is `LuaModule`, so it needs to implement its pure virtual functions. Second, we want it to resemble what we implemented in the factory exercise, which means we have a pretty good idea of what to return in the `luaRegs` virtual function implementation.

Let us get started. Add a new file named `LuaModuleExporter.hpp` and define the `LuaModuleExporter` class, as follows:

```
template <typename T>
class LuaModuleExporter final : public LuaModule
{
public:
    LuaModuleExporter(
        const LuaModuleExporter &) = delete;
    ~LuaModuleExporter() = default;

    static LuaModuleExporter<T> make()
    {
        return LuaModuleExporter<T>();
    }

private:
    LuaModuleExporter() {}
};
```

This makes the exporter a final class and prevents it from being copy-constructed. Because the purpose of the exporter is to provide LuaModule and we do not have logic for it to get passed around by value, adding some restrictions can prevent bugs in the future. We achieve this by assigning the `delete` keyword to the copy constructor. We also want to control the object creation, so we make the constructor private. This has another effect – you cannot use `new operator` to create instances of the class.

Now, add the implementation for LuaModule as follows:

```
class LuaModuleExporter final : public LuaModule
{
public:
    const std::string &luaName() const override
    {
        return name;
    }

    const std::vector<luaL_Reg> &luaRegs() const override
    {
        return factoryRegs;
    }

private:
    const std::string name = "TODO";
    const std::vector<luaL_Reg> factoryRegs = {
        {"new", luaNew},
        {NULL, NULL}};

    static int luaNew(lua_State *L)
    {
        return 0;
    }
};
```

This is straightforward. On the Lua module level, we only want to export one function to create concrete objects. So, we will only register `luaNew`. The name of the module needs to be passed in. We will find a way when we implement the details.

Thus, we have a stub for our exporter. This is a system-level design contract. Now, let us write the test code to see how it should be used.

Preparing the C++ test code

In `main.cpp`, write the `main` function as follows:

```
int main()
{
```

```
    auto listener = std::make_unique<
        LoggingLuaExecutorListener>();
    auto lua = std::make_unique<LuaExecutor>(*listener);

    auto module = LuaModuleExporter<
        Destinations>::make();
    lua->registerModule(module);

    lua->executeFile("script.lua");
    return 0;
}
```

Compared with the previous chapter, the only difference is how `LuaModule` is created. Now, it is created with `LuaModuleExporter<Destinations>::make()`.

At this point, the project should compile. When you run it, it should not crash on the C++ side; although, at this stage, it will not be able to do anything meaningful, and you should see an error message from Lua.

Now, we will see what Lua code we need.

Preparing the Lua test script

Write `script.lua` exactly as follows:

```
dst = Destinations.new()
dst:wish("London", "Paris", "Amsterdam")
dst:went("Paris")
print("Visited:", dst:list_visited())
print("Unvisited:", dst:list_unvisited())
```

We used this code snippet in the previous chapter. This will help to validate whether we will get the same result later in this chapter.

Next, let us start to make the exporter work.

Defining LuaModuleDef

First, we need to provide the name of the module and then the `__index` metatable. Finally, we need to provide a name for the metatable. Recall that in `Destinations.cc`, the name of the metatable is hardcoded as follows:

```
const std::string METATABLE_NAME(
    "Destinations.Metatable");
```

Now, this needs to be passed to the exporter. Let us define a structure for the aforementioned three pieces of information. In LuaModule.h, add the following declaration:

```
template <typename T>
struct LuaModuleDef
{
    const std::string moduleName;
    const std::vector<luaL_Reg> moduleRegs;

    const std::string metatableName() const
    {
        return std::string(moduleName)
            .append(".Metatable");
    }
};
```

This defines moduleName and moduleRegs. The metatable name is based on the module name, with ".Metatable" appended to it.

Note that this structure is also templated. This indicates that a definition is for a certain C++ type. We will use the template later in this chapter.

Now, we can pass this structure to the exporter.

Using LuaModuleDef

In LuaModuleExporter.hpp, accept an instance of LuaModuleDef during exporter creation. Rewrite the related code as follows:

```
class LuaModuleExporter final : public LuaModule
{
public:
    static LuaModuleExporter<T> make(
        const LuaModuleDef<T> &luaModuleDef)
    {
        return LuaModuleExporter<T>(luaModuleDef);
    }

    const std::string &luaName() const override
    {
        return luaModuleDef.moduleName;
    }

private:
    LuaModuleExporter(
```

```
            const LuaModuleDef<T> &luaModuleDef)
            : luaModuleDef(luaModuleDef) {}

        const LuaModuleDef<T> luaModuleDef;
};
```

The changes are as follows:

- We added a private member variable, `luaModuleDef`
- We added an argument of type `LuaModuleDef` to `make` and the private constructor
- We changed `luaName` to return `luaModuleDef.moduleName`
- We deleted the private member variable, `name`, introduced during stubbing

Now, we can define `LuaModuleDef` for the `Destinations` class.

In `Destinations.h`, delete the declaration for `DestinationsFactory` and add the following code:

```
struct DestinationsLuaModuleDef
{
    static LuaModuleDef<Destinations> def;
};
```

In `Destinations.cpp`, delete all implementations for `DestinationsFactory` and add the following code after the anonymous namespace:

```
LuaModuleDef DestinationsLuaModuleDef::def =
    LuaModuleDef<Destinations>{
        "Destinations",
        {{"wish", luaWish},
         {"went", luaWent},
         {"list_visited", luaListVisited},
         {"list_unvisited", luaListUnvisited},
         {NULL, NULL}},
    };
```

Finally, in `main.cpp`, change the module creation code to the following statement:

```
auto module = LuaModuleExporter<Destinations>::make(
    DestinationsLuaModuleDef::def);
```

This pumps `LuaModuleDef` for the `Destinations` class into the exporter. Make sure that the project can compile.

Now, we will fill in the rest of the missing pieces to make the exporter really work.

Re-implementing luaNew

Because we will store LuaModuleDef in LuaModuleExporter, to access it, we need to find the instance of LuaModuleExporter. Let us first implement a helper function for this.

Since the exporter is also LuaModule, it already has an upvalue mechanism, implemented in *Chapter 6*. LuaModule::pushLuaUpvalues will push the pointer to the LuaModule instance as an upvalue. To retrieve it, we can add the following function:

```
class LuaModuleExporter final : public LuaModule
{
private:
    static LuaModuleExporter<T> *getExporter(
        lua_State *L)
    {
        return reinterpret_cast<LuaModuleExporter<T> *>(
            lua_touserdata(L, lua_upvalueindex(1)));
    }
};
```

This is the same as the getObj function in *Chapter 6*, but now, it is a static member function.

With a way to access the exporter instance from the static member functions, we can write LuaModuleExporter::luaNew as follows:

```
static int luaNew(lua_State *L)
{
    auto luaModuleDef = getExporter(L)->luaModuleDef;
    T *obj = new T();

    T **userdata = reinterpret_cast<T **>(
        lua_newuserdatauv(L, sizeof(obj), 0));
    *userdata = obj;

    auto metatableName = luaModuleDef.metatableName();
    int type = luaL_getmetatable(
        L, metatableName.c_str());
    if (type == LUA_TNIL)
    {
        lua_pop(L, 1);

        luaL_newmetatable(L, metatableName.c_str());
        lua_pushvalue(L, -1);
        lua_setfield(L, -2, "__index");
        luaL_setfuncs(
```

```
            L, luaModuleDef.moduleRegs.data(), 0);

        lua_pushcfunction(L, luaDelete);
        lua_setfield(L, -2, "__gc");
    }
    lua_setmetatable(L, 1);

    return 1;
}
```

This is actually copied from `Destinations.cc`. The changes, besides using `T` typename instead of the hardcoded class name, are highlighted in the preceding code. You can see that they are all about pumping `LuaModuleDef`.

If you have forgotten how `luaNew` works, you can check the previous chapter, where there are figures to show how the Lua stack changes as well.

Finally, let us implement the stub for `LuaModuleExporter::luaDelete` as follows:

```
static int luaDelete(lua_State *L)
{
    T *obj = *reinterpret_cast<T **>(
        lua_touserdata(L, 1));
    delete obj;
    return 0;
}
```

`luaDelete` is registered as the `__gc` metamethod in `luaNew`.

> **Do you remember?**
>
> As explained in the previous chapter, we set `luaDelete` as the finalizer for the user data created in `luaNew`. During the Lua garbage collection process, the finalizer will be called, with an argument as the user data reference.

You can also delete REGS, FACTORY_REGS, luaNew, and `luaDelete` in `Destinations.cc`. They are not used anymore.

Now, we can test the exporter. Execute the project. If you have done everything correctly, you should see the following output:

```
Destinations instance created: 0x12a704170
[Lua] Visited: Paris
[Lua] Unvisited: Amsterdam London
Destinations instance destroyed: 0x12a704170
```

We have not really changed the test code from the previous chapter, except for how the Destinations class gets exported to Lua.

If you have encountered any errors, do not feel discouraged. This is the most complex chapter in this book, and we need to implement code correctly in multiple files to make it work. Trace back your steps and fix the error. You can do it! Also, in GitHub, there are multiple checkpoint projects for this chapter, which you can refer to. As previously mentioned, we will not automate the generation of the lua_CFunction wrappers. Generalization also needs a limit.

But, let us check how general our implementation is.

Are you flexible enough?

To answer this question, let us rewrite script.lua as follows:

```
dst = Destinations.new("Shanghai", "Tokyo")
dst:wish("London", "Paris", "Amsterdam")
dst:went("Paris")
print("Visited:", dst:list_visited())
print("Unvisited:", dst:list_unvisited())
```

Yes, the new requirement is that, in the Lua code, when creating the Destinations objects, we can provide an initial list of unvisited places.

This means that we need to support parameterized object creation.

Can our exporter support this? This should be a common use case.

Now is a good time to ponder over life, get a cup of coffee, or whatever. We are almost near the end of *Part 3* of this book.

If you recall, our object creation code is as follows:

```
static int luaNew(lua_State *L)
{
    ...
    T *obj = new T();
    ...
}
```

As a seasoned C++ programmer, you may think that, because std::make_unique<T> can forward its arguments to the constructor of T, there must be a way to make LuaModuleExporter<T>::make do the same. Right, but the magic of std::make_unique<T> is at C++ compile time. So, how would you handle that when the arguments are passed in Lua code after the C++ code has been compiled?

Worry not. Let us explore the **factory method** design pattern. A factory method is a contract defined as a method or an interface to create and return an object. However, how the object is created is not important and not part of the contract.

To see how it works, let us implement one for `LuaModuleDef`. Add another member variable, named `createInstance`, as follows:

```
struct LuaModuleDef
{
    const std::function<T *(lua_State *)>
    createInstance =
    [](lua_State *) -> T* { return new T(); };
};
```

This is a bit of advanced C++ usage. Therefore, it is important that you take the following into account:

- `createInstance` is declared as a member variable but not as a member function. This is because you can simply assign the member variable a different value during object construction to achieve a different behavior, but with a member function, you need to create a subclass to override the behavior. *We should prefer composition over inheritance whenever we can.*

- `createInstance` is of type `std::function`. With this type, you can use the variable as if it were a function. If you are more familiar with Lua in this regard, you'll understand that a named Lua function is also a variable. Here, we want to achieve the same effect. `T *(lua_State *)` is the type of the function. It means that the function expects one argument of type `lua_State*` and will return a pointer to type T. You can check the C++ reference manual to learn more about `std::function`.

- Then, we provide a default implementation as a C++ lambda. This lambda simply creates an instance in the heap without any constructor parameter.

To use this factory method, change `LuaModuleExporter::luaNew`, as follows:

```
static int luaNew(lua_State *L)
{
    . . .
    T *obj = luaModuleDef.createInstance(L);
    . . .
}
```

We have changed from new `T()` to `luaModuleDef.createInstance(L)`, and it still does the same thing.

However, note that we no longer create the object in `LuaModuleExporter`.

Finally, to answer the question, yes, we are flexible enough.

> **On modern C++**
>
> In 1998, C++ was standardized for the first time as C++98. It saw little change until 2011, with C++11. Since then, C++ has quickly adopted modern programming techniques in language specification. Lambdas and `std::function` are just two of the many examples. If you know some other languages (for example, Java), you can make some analogies (lamdba and functional interface), although the syntaxes are different. I implemented LuaModuleDef this way instead of using a more traditional method to show you some examples of modern C++ features. This is the future, and I encourage you to explore modern C++ in more detail. People working with Java, Kotlin, and Swift use such techniques by default. You can play an important role here by adopting these new techniques and helping C++ to catch up.

In `Destinations.cc`, change the `LuaModuleDef` instance as follows:

```
LuaModuleDef DestinationsLuaModuleDef::def =
LuaModuleDef<Destinations>{
    "Destinations",
    {{"wish", luaWish},
    ...
    {NULL, NULL}},
    [](lua_State *L) -> Destinations *
    {
        Destinations *obj = new Destinations();
        std::vector<std::string> places;
        int nArgs = lua_gettop(L);
        for (int i = 1; i <= nArgs; i++)
        {
            places.push_back(lua_tostring(L, i));
        }
        obj->wish(places);
        return obj;
    },
};
```

This initializes the `createInstance` field with the provided lambda, rather than the default lambda. The new lambda does similar things to the `luaWish` wrapper. The beauty of this is that you have full control over this lambda. You can create another constructor for the `Destinations` class and simply invoke the new constructor.

We can test the project with the new Lua script. You should see the following output:

```
Destinations instance created: 0x142004170
[Lua] Visited: Paris
[Lua] Unvisited: Amsterdam London Shanghai Tokyo
Destinations instance destroyed: 0x142004170
```

As you can see, `Shanghai` and `Tokyo` have been added to the unvisited list.

> **Even further design improvement**
>
> We are creating objects in `LuaModuleDef` but destroying them in `LuaModuleExporter`, and our use case does not involve transferring object ownership. For a better design, the same class should destroy the objects it creates, which we will implement in the next chapter.

This time, for real, we have finished.

Summary

In this chapter, we implemented a general C++ module exporter, mainly for the object creation part. This ensures that you can implement complex object creation logic once and reuse it with many C++ classes. Also, this chapter marks the end of *Part 3, How to Call C++ from Lua*.

In the following chapter, we will recap the different communication mechanisms between Lua and C++ and explore them further.

Exercise

This is an open exercise. You can write a new C++ class, or find one from your work in the past, and then export it to Lua with `LuaModuleExporter`. Try to provide an interesting `createInstance` implementation as well.

Part 4 –
Advanced Topics

By this point of the book, you will have learned all the common mechanisms to integrate Lua with C++.

In this part, you will recap what you have learned, which will also serve as a source for quick reference. You will also learn how to implement a standalone C++ module that can be loaded by Lua, as a dynamic loadable library. Then, you will learn some advanced memory management techniques and how to implement multithreading with Lua.

This part comprises the following chapters:

- *Chapter 9, Recapping Lua-C++ Communication Mechanisms*
- *Chapter 10, Managing Resources*
- *Chapter 11, Multithreading with Lua*

9

Recapping Lua-C++ Communication Mechanisms

In *Part 2* of this book, we learned how to call Lua from C++. In *Part 3*, we learned how to call C++ from Lua. In the course of this book, we have explored many examples, some of which depend on advanced C++ techniques.

This chapter will summarize all the communication mechanisms between Lua and C++, stripping away most of the C++ details. We will also dig deeper into some of the topics that we have not demonstrated in the examples yet.

You can use this chapter to recap what you have learned. For each topic, we will list some important Lua library functions. You can check the Lua reference manual for more related functions.

In the future, as you progress in your programming journey, you might adopt different C++ techniques in your projects. In such cases, this chapter will be a useful source for quick reference.

We will cover the following topics:

- The stack
- Calling Lua from C++
- Calling C++ from Lua
- Implementing standalone C++ modules
- Storing state in Lua
- Userdata

Technical requirements

You can access the source code for this chapter at `https://github.com/PacktPublishing/Integrate-Lua-with-CPP/tree/main/Chapter09`.

You can access the Lua reference manual and develop the habit of frequently checking for API details at `https://www.lua.org/manual/5.4/`.

The stack

The Lua stack can serve two purposes:

- *Exchange data between C++ and Lua.* Passing function arguments and retrieving function return values fit into this usage.
- *Keep intermediate results.* For example, we can keep a table reference in the stack until we are done with the table; we can push some values onto the stack and then pop and use them as upvalues.

The Lua stack comes in two forms:

- *The public stack that comes with the Lua state.* Once a Lua state is created via `luaL_newstate` or `lua_newstate`, you can pass the state around and the same Lua stack is accessible to all functions that can access the Lua state.
- *The private stack for each* `lua_CFunction` *call.* The stack is only accessible to a function call. Calling the same `lua_CFunction` multiple times will not share the same stack. So, the stack that is passed to a `lua_CFunction` call is private to the function call.

Pushing onto the stack

You can use `lua_pushXXX` functions to push a value or an object reference onto the stack – for example, `lua_pushstring`.

Check the Lua reference manual for a list of such functions.

Querying the stack

You can use `lua_isXXX` functions to check if a given stack position holds an item of a certain type.

You can use `lua_toXXX` functions to convert a given stack position into a certain type. Those functions will always succeed, although the resulting values might be a surprise if the stack position is holding an item of a different type.

You can check the Lua reference manual for a list of such functions.

Other stack operations

There are some other frequently used stack operations.

Ensuring stack size

The Lua stack is created with a predefined size that should be big enough for most operations. If you need to push a lot of items onto the stack, you can ensure that the stack size can meet your needs by calling the following function:

```
int lua_checkstack (lua_State *L, int n);
```

n is the required size.

Counting items

To check the number of items in the stack, use lua_gettop. The return value is the count.

Resetting the stack top

To set the top of the stack to a certain index, use the lua_settop function, whose declaration is as follows:

```
void lua_settop (lua_State *L, int index);
```

This can either clear some items from the top of the stack or pad the stack with nils. We can use it to clear temporary items from the stack efficiently, as can be seen in LuaModuleExporter::luaNew from our examples:

```
T *obj = luaModuleDef.createInstance(L);
lua_settop(L, 0);
```

In luaNew, we passed the Lua state, thus the Lua stack, to an external factory method. Because we do not know how the factory method will use the Lua stack, we cleared the stack after the factory method returned to get rid of any possible side effects.

Copying another item

If an item is already in the stack, you can push a copy of it onto the top of the stack quickly by calling this function:

```
void lua_pushvalue (lua_State *L, int index);
```

This can save you some trouble if the value or object backing the item is hard to get.

Some other stack operations are supported by the Lua library, but they are used infrequently to achieve complex effects. You can check them out in the reference manual.

Calling Lua from C++

To call Lua code from C++, we can use `lua_pcall`, which is declared as follows:

```
int lua_pcall(
    lua_State *L, int nargs, int nresults, int msgh);
```

This will call a Lua callable, which can be a function or a chunk. You can push the Lua function to be called onto the stack, or compile a file or a string into a chunk and then place it onto the stack. `nargs` is the number of arguments for the callable. The arguments are pushed onto the stack above the callable. `nresults` is the count of return values the callable would return. Use LUA_MULTRET to indicate that you expect a variable count of return values. `msgh` is the stack index for an error message handler.

`lua_pcall` calls the callable in *protected mode*, which means that any error that may have occurred in the call chain is not propagated. Instead, an error status code is returned from `lua_pcall`.

In the `LuaExecutor` class that we have implemented, you can find many examples of calling Lua from C++.

In the Lua reference manual, you can find other library functions similar to `lua_pcall`, although `lua_pcall` is the most frequently used one.

Calling C++ from Lua

To call C++ code from Lua, the C++ code needs to be exported via a `lua_CFunction` implementation, which is defined as follows:

```
typedef int (*lua_CFunction) (lua_State *L);
```

For example, in `LuaExecutor`, we implemented a function:

```
int luaGetExecutorVersionCode(lua_State *L)
{
    lua_pushinteger(L, LuaExecutor::versionCode);
    return 1;
}
```

This returns a single integer value to the Lua code. A simple way to export this function to the global table can be implemented as follows:

```
void registerHostFunctions(lua_State *L)
{
    lua_pushcfunction(L, luaGetExecutorVersionCode);
    lua_setglobal(L, "host_version");
}
```

You can use `lua_pushcfunction` to push `lua_CFunction` onto the stack and then assign it to a variable of your choice.

However, more than likely, you should export a group of functions as a module.

Exporting C++ modules

To export a C++ module, you simply need to export a table of functions to Lua. In `LuaExecutor`, we have implemented it as follows:

```
void LuaExecutor::registerModule(LuaModule &module)
{
    lua_createtable(L, 0, module.luaRegs().size() - 1);
    int nUpvalues = module.pushLuaUpvalues(L);
    luaL_setfuncs(L, module.luaRegs().data(), nUpvalues);
    lua_setglobal(L, module.luaName().c_str());
}
```

The process is to first create a table and push the reference onto the stack with `lua_createtable`. Then, you can push *shared upvalues* (we will recap upvalues later in this chapter), and finally add the list of functions to the table with `luaL_setfuncs`.

If you do not need upvalues, there is a shortcut that you can use:

```
void luaL_newlib (lua_State *L, const luaL_Reg l[]);
```

Both `luaL_newlib` and `luaL_setfuncs` take a list of the following structure to describe the functions:

```
typedef struct luaL_Reg {
    const char *name;
    lua_CFunction func;
} luaL_Reg;
```

The structure provides `lua_CFunction` with a `name` value, which is used as the table entry key.

Implementing standalone C++ modules

So far in this book, we have only explicitly registered C++ modules to Lua in C++ code. However, there is another way to provide a C++ module to Lua.

You can produce a shared library for a module and place it in Lua's search path. When the Lua code *requires* the module, Lua will load the shared library automatically.

By reusing our `Destinations` class, this is simple to implement. Create a file named `DestinationsModule.cpp` and fill it exactly as follows:

```cpp
#include "Destinations.h"
#include "LuaModuleExporter.hpp"
#include <lua.hpp>

namespace {
    LuaModuleExporter module =
        LuaModuleExporter<Destinations>::make(
            DestinationsLuaModuleDef::def);
}

extern "C" {
int luaopen_destinations(lua_State *L)
{
    lua_createtable(L, 0, module.luaRegs().size() - 1);
    int nUpvalues = module.pushLuaUpvalues(L);
    luaL_setfuncs(L, module.luaRegs().data(), nUpvalues);
    return 1;
}
}
```

The module that's been implemented is called `destinations`. The code-level contract Lua requires is as follows:

- You need to provide `lua_CFunction`, whose name must begin with `luaopen_`, and then have the module name appended
- `lua_CFunction` needs to leave what it creates in the stack

The code for `luaopen_destinations` is almost identical to that of `LuaExecutor::registerModule`, which we explained in the previous section. The only difference is that we have left the table reference in the stack because the Lua `require` function will pop it.

---

extern "C"

By default, the C++ compiler will mangle the C++ function name. This means that after the function is compiled, the function will have a symbol name more complex than what it declares to be in the source code. To prevent this from happening, you can place the function declaration inside an `extern "C"` block. Otherwise, Lua won't be able to find the function as the contract is broken after compilation.

Compiling the standalone module

To compile the shared library, add the following lines to your `Makefile`:

```
DESTINATIONS_O = Destinations.o DestinationsModule.o
DESTINATIONS_SO = destinations.so
destinations: ${DESTINATIONS_O}
    $(CXX) $(CXXFLAGS) $(CPPFLAGS) $(LDFLAGS) -shared \
        -o $(DESTINATIONS_SO) ${DESTINATIONS_O} -llua
```

In a Terminal, execute `make destinations` to create the shared library. You will get a file named `destinations.so`, which is the binary file Lua will load.

Testing the standalone module

To test the standalone module, in the folder where `destinations.so` resides, start a Lua interactive interpreter and execute the following statements:

```
Chapter09 % ../lua/src/lua
Lua 5.4.6 Copyright (C) 1994-2023 Lua.org, PUC-Rio
> Destinations = require "destinations"
> dst = Destinations.new("Shanghai", "Tokyo")
Destinations instance created: 0x155a04210
> dst:wish("London", "Paris", "Amsterdam")
> dst:went("Paris")
> print("Visited:", dst:list_visited())
Visited: Paris
> print("Unvisited:", dst:list_unvisited())
Unvisited: Amsterdam London Shanghai Tokyo
> os.exit()
Destinations instance destroyed: 0x155a04210
```

The most important statement is the `require` statement. This loads `destinations.so` and assigns the module to the `Destinations` global variable.

We started the Lua interactive interpreter in the same folder where the module binary resides because `require` will search the current working directory for modules. Alternatively, you can put the library in a system search path. You can check the reference manual to learn more about `require` and its behaviors.

A standalone C++ module is useful when you need to reuse the module in the binary form across multiple projects or enforce code isolation on the C++ side, but this is just a design choice.

Storing state in Lua

There are two ways to store state in Lua for `lua_CFunction`: *upvalues* and *the registry*. Let's recap them and dig deeper into upvalues.

Upvalues

To introduce the complete definition for upvalues, we need to introduce **Lua C closures** at the same time. To quote the Lua reference manual:

> *When a C function is created, it is possible to associate some values with it, thus creating a C closure; these values are called upvalues and are accessible to the function whenever it is called.*

To put it simply, the closure is still our old friend `lua_CFunction`. When you associate some values with it, it becomes a closure, and the values become upvalues.

It is important to note that Lua C closures and upvalues are inseparable.

To create a closure, use the following library function:

```
void lua_pushcclosure(
    lua_State *L, lua_CFunction fn, int n);
```

This creates a closure from `lua_CFunction` and associates n values with it.

To see it in action, let's solve the design problem from the previous chapter:

> *We are creating objects in* `LuaModuleDef` *but destroying them in* `LuaModuleExporter`. *For a better design, the same class should destroy the objects it creates.*

Implementing a Lua C closure

The following feature is a continuation of the previous chapter. You can revisit the previous chapter to get a better understanding if you need to.

To do this, we can implement a `destroyInstance` member variable for LuaModuleDef, as follows:

```
struct LuaModuleDef
{
    ...
    const std::function<void(T *)> destroyInstance =
        [](T *obj) { delete obj; };
    ...
};
```

Now, objects will be created and destroyed in the same LuaModuleDef entity. To use destroyInstance, modify LuaModuleExporter::luaDelete, as follows:

```
static int luaDelete(lua_State *L)
{
    auto luaModuleDef = getExporter(L)->luaModuleDef;
    T *obj = *reinterpret_cast<T **>(
        lua_touserdata(L, 1));
    luaModuleDef.destroyInstance(obj);
    return 0;
}
```

Recall that getExporter is used to retrieve the first upvalue, which is a pointer to the exporter:

```
static LuaModuleExporter<T> *getExporter(lua_State *L)
{
    return reinterpret_cast<LuaModuleExporter<T> *>(
        lua_touserdata(L, lua_upvalueindex(1)));
}
```

This works for luaNew because LuaModuleExporter is inherited from LuaModule, which pushes this as an upvalue in its default implementation:

```
class LuaModule
{
public:
    virtual int pushLuaUpvalues(lua_State *L)
    {
        lua_pushlightuserdata(L, this);
        return 1;
    }
};
```

Then, the pushed upvalue is used as shared upvalues for all exported functions in LuaExecutor::registerModule:

```
void LuaExecutor::registerModule(LuaModule &module)
{
    lua_createtable(L, 0, module.luaRegs().size() - 1);
    int nUpvalues = module.pushLuaUpvalues(L);
    luaL_setfuncs(L, module.luaRegs().data(), nUpvalues);
    lua_setglobal(L, module.luaName().c_str());
}
```

Shared upvalues are pushed onto the stack once and get associated with all the functions provided to luaL_setfuncs.

> **Shared upvalues are not really shared**
>
> The so-called shared upvalues are copied for each function during setup. Afterward, the functions access their own copies of the upvalues. In the Lua reference manual, these are called shared upvalues because they are only pushed onto the stack once and used for all functions to be registered, which is only relevant to API invocations. I think this term is misleading. You should think of these as just plain upvalues.

However, `getExporter` will not work for `luaDelete` because `luaDelete` is not an exported function and is not passed to `luaL_setfuncs`. To support `luaDelete`, modify `luaNew`, as follows:

```
static int luaNew(lua_State *L)
{
    auto exporter = getExporter(L);
    auto luaModuleDef = exporter->luaModuleDef;

    ...

    if (type == LUA_TNIL)
    {
        ...

        lua_pushlightuserdata(L, exporter);
        lua_pushcclosure(L, luaDelete, 1);
        lua_setfield(L, -2, "__gc");
    }
    ...
}
```

We only need to push `exporter` as an upvalue for `luaDelete` and make `luaDelete` a closure.

Now, `LuaModuleExporter` has a better design as it delegates both object construction and object destruction to `LuaModuleDef`. Also, it utilizes both upvalues (for `luaDelete`) and shared upvalues (for `luaNew`) at the same time in the `getExporter` helper function. This shows that shared upvalues are no different from upvalues after they are set up.

The registry

The registry is a predefined Lua table that is only accessible to C/C++ code. For a Lua state, the registry is shared for all C/C++ functions, so table key names should be selected carefully to avoid collision.

Notably, by convention, *full userdata* places its metatable in the registry via `luaL_newmetatable`.

Put simply, the registry is a Lua table that the Lua language treats specially and provides a few helper functions for.

Userdata

The Lua userdata can be categorized into *light userdata* and *full userdata*.

It is important to note that they are different things. In the Lua library, conventionally, light userdata is named lightuserdata, while full userdata is named userdata.

Light userdata

Light userdata represents a C/C++ pointer. It is a value type and the value is passed around. You push a pointer in C/C++ code onto the stack with `lua_pushlightuserdata`. You cannot create light userdata with the Lua library.

Full userdata

Full userdata is a raw memory area allocated by the Lua library with a call to `lua_newuserdatauv`. It is an object type and only its reference is passed around.

Because full userdata is created by Lua in the heap, Lua garbage collection comes into the picture. On the C++ side, you can provide a *finalizer* by providing the `__gc` metamethod.

For a complete example of how to utilize full userdata to access C++ objects in Lua, check `LuaModuleExporter`.

Summary

In this chapter, we briefly recapped all communication mechanisms between Lua and C++. This should have sufficiently reinforced your learnings so far.

We also learned how to produce a standalone C++ module as a shared library. This opens new ways for you to organize your projects.

In the next chapter, we will talk more about resource management.

10

Managing Resources

In the previous chapter, we recapped the communication mechanisms between Lua and C++. In this chapter, we will learn more about managing resources. Resources can be anything an object uses, such as memory, files, or network sockets.

We will cover the following topics:

- Customizing Lua memory allocation
- Delegating C++ object memory allocation to Lua
- What is RAII?

Technical requirements

We will use the source code for *Chapter 9* as a base to develop the examples in this chapter. Make sure you can access the source code for this book: `https://github.com/PacktPublishing/Integrate-Lua-with-CPP/tree/main/Chapter10`.

Customizing Lua memory allocation

In the Lua runtime, memory is allocated, reallocated, or deallocated in the heap in the following situations:

- **Memory is allocated**: This happens when an object is created. Lua needs to allocate a piece of memory to hold it.
- **Memory is reallocated**: This happens when the size of an object needs to be changed – for example, adding entries to a table when the table has no more pre-allocated space.
- **Memory is deallocated**: This happens during garbage collection when the object is no longer needed.

In most situations, you do not need to be concerned about this. But sometimes, it is helpful to get an insight into, or customize, the Lua memory allocation. Here are some examples:

- You need to analyze the memory footprint of your Lua objects to find optimization opportunities.
- You need to customize where the memory is allocated. For example, to increase runtime efficiency, you may have a memory pool and you can simply have Lua use it without allocating new memory regions in the heap every time.

In this section, we will see how we can customize Lua's memory allocation by providing a memory allocation function.

What is the Lua memory allocation function?

Lua provides a simple way to customize memory allocation. When you create a Lua state, you can provide a memory allocation function so that whenever Lua needs to manage memory, it calls the function you provided.

The memory allocation function is defined as the following type:

```
typedef void * (*lua_Alloc) (void *ud,
                             void *ptr,
                             size_t osize,
                             size_t nsize);
```

The function returns a pointer to the newly allocated memory, or NULL if the call is to deallocate a piece of memory. Its arguments are explained as follows:

- ud is the pointer to user-defined data for the Lua state. You can use the same memory allocation function with many Lua states. In such cases, you can use ud to identify each Lua state. Lua treats this transparently.
- ptr is the pointer to the memory to be reallocated or deallocated. If it is NULL, the call to the memory allocator is to allocate a new piece of memory.
- osize is the original size for a previously allocated memory pointed by ptr. If ptr is NULL, osize has a special meaning – the type of the Lua object that is being allocated for, which can be LUA_TSTRING, LUA_TTABLE, and so on.
- nsize is the size for the memory to be allocated or reallocated. If nsize is 0, the memory is to be deallocated.

To register your memory allocation function, you can use lua_newstate to create the Lua state, which is declared as follows:

```
lua_State *lua_newstate (lua_Alloc f, void *ud)
```

With this, you provide both the memory allocation function and the user data for the Lua state to be created. Note that you can provide NULL to ud, and this user data is a C++ side object, not the Lua user data.

Next, we'll implement a memory allocation function.

Implementing a memory allocation function

We will extend LuaExecutor to practice implementing a memory allocation function. When we create an executor, we want to pass a flag to indicate whether we should use our own memory allocation function.

You can start this work based on the source code for *Chapter 9*. In LuaExecutor.h, change the constructor, as follows:

```
class LuaExecutor
{
public:
    LuaExecutor(const LuaExecutorListener &listener,
                bool overrideAllocator = false);
};
```

We added another Boolean argument named overrideAllocator for the constructor. We also provided a default value as false because in most cases, we do not need to override the Lua memory allocator.

In LuaExecutor.cc, implement our memory allocation function in a new anonymous namespace, as follows:

```
namespace
{
void *luaAlloc(
    void *ud, void *ptr, size_t osize, size_t nsize)
{
    (void)ud;
    std::cout << "[luaAlloc] ptr=" << std::hex << ptr
             << std::dec << ", osize=" << osize
             << ", nsize=" << nsize;
    void *newPtr = NULL;
    if (nsize == 0)
    {
        free(ptr);
    }
    else
    {
```

```
        newPtr = realloc(ptr, nsize);
    }
    std::cout << std::dec << ", newPtr=" << newPtr
              << std::endl;
    return newPtr;
}
}
```

`luaAlloc` relies on the standard `realloc` and `free` C functions to allocate, reallocate, and deallocate memory. This is exactly what the default Lua allocator does. But we also log the arguments and the return value to get more insight into the memory usage.

To use `luaAlloc`, in `LuaExecutor.cc`, modify the constructor, as follows:

```
LuaExecutor::LuaExecutor(
    const LuaExecutorListener &listener,
    bool overrideAllocator)
    : L(overrideAllocator ? lua_newstate(luaAlloc, NULL)
                          : luaL_newstate()),
      listener(listener)
{ ... }
```

Here, we check if `overrideAllocator` is `true`. If it is, we use our memory allocation function by calling `lua_newstate`. If it is not, we use the default allocator by calling `luaL_newstate`.

Now, let's test our allocator.

Testing it out

Rewrite `main.cpp`, as follows:

```
#include "LuaExecutor.h"
#include "LoggingLuaExecutorListener.h"
#include "LuaModuleExporter.hpp"
#include "Destinations.h"

int main()
{
    auto listener = std::make_unique<
        LoggingLuaExecutorListener>();
    auto lua = std::make_unique<LuaExecutor>(
        *listener, true);
    auto module = LuaModuleExporter<Destinations>::make(
        DestinationsLuaModuleDef::def);
    lua->registerModule(module);
```

```
    lua->executeFile("script.lua");
    return 0;
}
```

The test code creates a Lua executor, registers the Destinations module, and executes script. lua. This is similar to what we did in the previous chapters. The only thing to note is that we are setting overrideAllocator to true when creating the LuaExecutor instance.

Rewrite script.lua, as follows:

```
print("======script begin======")
dst = Destinations.new()
dst:wish("London", "Paris", "Amsterdam")
dst:went("Paris")
print("Visited:", dst:list_visited())
print("Unvisited:", dst:list_unvisited())
print("======script end======")
```

The script creates an object of the Destinations type and tests its member functions. This, again, is similar to what we did in the previous chapters.

We also print out markers to mark when the script starts and finishes execution. This helps us locate things of interest because the customized memory allocation function will be quite verbose.

Compile and execute the project. You should get an output similar to the following:

```
...
[Lua] ======script begin======
[luaAlloc] ptr=0x0, osize=7, nsize=56, newPtr=0x14e7060c0
Destinations instance created: 0x14e7060e0
[luaAlloc] ptr=0x0, osize=4, nsize=47, newPtr=0x14e706100
[luaAlloc] ptr=0x0, osize=5, nsize=56, newPtr=0x14e706130
[luaAlloc] ptr=0x0, osize=0, nsize=48, newPtr=0x14e706170
[luaAlloc] ptr=0x0, osize=0, nsize=96, newPtr=0x14e7061a0
[luaAlloc] ptr=0x14e706170, osize=48, nsize=0, newPtr=0x0
...
[Lua] Visited: Paris
[Lua] Unvisited: Amsterdam London
[Lua] ======script end======
Destinations instance destroyed: 0x14e7060e0
...
```

The two lines highlighted here are the allocation and deallocation of the object at the 0x14e706170 address. You will also see a lot of unrelated memory allocation outputs because Lua will also use the customized memory allocation function to manage the memory of its internal states.

Although this customized memory allocation function is not very complex, you can extend what you have learned to change how memory is managed. This is useful for runtime optimization or resource-restricted systems.

In the next section, we will explore a higher-level scenario – *how to make Lua allocate memory for C++ objects*.

Delegating C++ object memory allocation to Lua

So far, we have been creating C++ objects in C++ and making Lua store its pointer in userdata. This was done in `LuaModuleExporter::luaNew`, as follows:

```
static int luaNew(lua_State *L)
{
    ...
    T **userdata = reinterpret_cast<T **>(
        lua_newuserdatauv(L, sizeof(T *), 0));
    T *obj = luaModuleDef.createInstance(L, nullptr);
    *userdata = obj;
    ...
}
```

In this case, the Lua userdata only stores a pointer. As you may recall, Lua userdata can represent a much larger piece of memory, so you might be wondering if we can store the whole C++ object in userdata, instead of just the pointer. Yes, we can. Let's learn how to do it.

Using C++ placement new

In C++, the most common way to create an object is to call `new  T()`. This does two things:

- It creates a piece of memory to hold an object of the T type.
- It calls a constructor of the T type. In our example, we are calling the default constructor.

Similarly, the most common way to destroy an object is to call `delete  obj`. It also does two things:

- It calls the destructor of the T type. Here, `obj` is an object of the T type.
- Frees the memory that holds `obj`.

C++ also provides another *new expression* that only constructs an object by calling a constructor. It does not allocate memory for the object. Instead, you tell C++ where to place the object. This *new expression* is called **placement new**.

To use *placement new*, we need to provide the address to a piece of memory that has already been allocated. We can use it in the following way:

```
T* obj = new (addr) T();
```

We need to provide the address to the memory location between the new keyword and the constructor.

Now that we have found a way to decouple C++ memory allocation and object construction, let's extend our C++ module exporter to support delegating memory management to Lua.

Extending LuaModuleDef

We have implemented a C++ module exporting system in this book. It has two parts:

- LuaModuleExporter abstracts the module registration and implements the Lua finalizer for the module

- LuaModuleDef defines the module name, exported functions, and object construction and destruction

First, we will add the capability to use pre-allocated memory in LuaModuleDef.

In LuaModule.h, add a new member variable named isManagingMemory, as follows:

```
struct LuaModuleDef
{
    const bool isManagingMemory;
};
```

When isManagingMemory is true, we indicate that the LuaModuleDef instance is managing memory allocation and deallocation. When isManagingMemory is false, we indicate that LuaModuleDef is not managing memory. In the latter case, LuaModuleExporter should make Lua manage memory, which we will implement after we have extended LuaModuleDef.

With the new flag added, modify createInstance, as follows:

```
const std::function<T *(lua_State *, void *)>
createInstance = [this](lua_State *, void *addr) -> T *
{
    if (isManagingMemory)
    {
        return new T();
    }
    else
    {
        return new (addr) T();
```

```
    }
  };
```

We added a new argument – `void *addr`. When the `LuaModuleDef` instance is managing memory, it allocates the memory with the normal *new operator*. When the instance is not managing memory, it uses the *placement new expression*, where `addr` is the address where the object should be constructed.

This implementation is the default implementation for `createInstance`. You can override it and call a non-default constructor when you create `LuaModuleDef` instances.

Next, we need to modify `destroyInstance` to support `isManagingMemory` as well. Change its default implementation, as follows:

```
const std::function<void(T *)>
destroyInstance = [this](T *obj)
{
    if (isManagingMemory)
    {
        delete obj;
    }
    else
    {
        obj->~T();
    }
};
```

When the `LuaModuleDef` instance is not managing memory, we simply call the object's destructor, `obj->~T()`, to destroy it.

> **Placement delete?**
>
> If you are wondering whether there is a *placement delete* to match the *placement new*, the answer is no. To destroy an object without deallocating its memory, you can simply call its destructor.

With `LuaModuleDef` ready to support two ways of memory management, next, we will extend `LuaModuleExporter`.

Extending LuaModuleExporter

Before we add support to delegate memory management for C++ objects to Lua, we will highlight the major architectural difference:

- When C++ allocates the object's memory, as we have done in this book till now, the Lua userdata holds a pointer to the address of the memory allocated
- When Lua allocates the object's memory as userdata, the userdata holds the actual C++ object

Let's start to extend LuaModuleExporter. We need to modify both luaNew and luaDelete so that they work with LuaModuleDef::isManagingMemory.

In LuaModuleExporter.hpp, change luaNew, as follows:

```
static int luaNew(lua_State *L)
{
    auto exporter = getExporter(L);
    auto luaModuleDef = exporter->luaModuleDef;

    if (luaModuleDef.isManagingMemory)
    {
        T **userdata = reinterpret_cast<T **>(
            lua_newuserdatauv(L, sizeof(T *), 0));
        T *obj = luaModuleDef.createInstance(L, nullptr);
        *userdata = obj;
    }
    else
    {
        T *userdata = reinterpret_cast<T *>(
            lua_newuserdatauv(L, sizeof(T), 0));
        luaModuleDef.createInstance(L, userdata);
    }

    lua_copy(L, -1, 1);
    lua_settop(L, 1);


    ...
}
```

The function is broken down into four blocks, separated by new lines. These blocks are as follows:

- The first two lines of code get the LuaModuleExporter instance and the LuaModuleDef instance. You can revisit *Chapter 8* to understand how getExporter works if you need to.

- The if clause creates the C++ module object and the Lua userdata. When luaModuleDef.isManagingMemory is true, the code that's executed is the same as that in *Chapter 8*. When it is false, the code creates a userdata with a size of sizeof(T) to hold the actual T instance. Note that, in this case, the type of the userdata is T*, and its address is passed to luaModuleDef.createInstance to use with the *placement new*.

- It copies the userdata to the bottom of the stack via `lua_copy(L, -1, 1)` and clears everything but the bottom of the stack via `lua_settop(L, 1)`. The object construction is delegated to `LuaModuleDef` to clear the stack of temporary items in case `LuaModuleDef` has pushed any. These two lines of code are an improved version compared to the code in *Chapter 8* to cover more cases and the different ways of object creation.

- The omitted code in the rest of the function is unchanged.

Finally, to complete the feature, modify `LuaModuleExporter::luaDelete`, as follows:

```
static int luaDelete(lua_State *L)
{
    auto luaModuleDef = getExporter(L)->luaModuleDef;
    T *obj = luaModuleDef.isManagingMemory
        ? *reinterpret_cast<T **>(lua_touserdata(L, 1))
        : reinterpret_cast<T *>(lua_touserdata(L, 1));
    luaModuleDef.destroyInstance(obj);
    return 0;
}
```

We need to change how to get the C++ module instance in the finalizer. The difference comes from whether the userdata is holding the actual object or a pointer to the object.

Next, let's test whether the mechanism to have Lua allocate the C++ object memory works.

Testing with the Destinations.cc module

We only need to tweak the code in `Destinations.cc` a little bit to support both memory allocation scenarios. Change `getObj`, as follows:

```
inline Destinations *getObj(lua_State *L)
{
    luaL_checkudata(L, 1, DestinationsLuaModuleDef::def
        .metatableName().c_str());
    if (DestinationsLuaModuleDef::def.isManagingMemory)
    {
        return *reinterpret_cast<Destinations **>(
            lua_touserdata(L, 1));
    }
    else
    {
        return reinterpret_cast<Destinations *>(
            lua_touserdata(L, 1));
    }
}
```

The change we've made here is similar to what we did to `LuaModuleExporter::luaDelete` so that it supports the different content the Lua userdata holds.

To choose to have Lua allocate the memory, change `DestinationsLuaModuleDef::def`, as follows:

```
LuaModuleDef DestinationsLuaModuleDef::def =
LuaModuleDef<Destinations>{
     "Destinations",
     {{"wish", luaWish},
      {"went", luaWent},
      {"list_visited", luaListVisited},
      {"list_unvisited", luaListUnvisited},
      {NULL, NULL}},
     false,
};
```

Here, we set `LuaModuleDef::isManagingMemory` to `false`.

Compile and execute the project. You should see the following output:

```
Chapter10 % ./executable
[Lua] ======script begin======
Destinations instance created: 0x135e0af10
[Lua] Visited: Paris
[Lua] Unvisited: Amsterdam London
[Lua] ======script end======
Destinations instance destroyed: 0x135e0af10
```

If you set `LuaModuleDef::isManagingMemory` to `true`, it should also work.

> **Who should manage the memory for C++ objects?**
>
> You can have either C++ or Lua manage the memory allocation for C++ objects. Managing memory in C++ can provide better control in complex projects. Managing memory in Lua can get rid of the double-pointer indirection. There are also psychological considerations. For some people coming from the C++ world, having Lua allocate C++ objects, especially when Lua is only one of the libraries used in the project, may seem to violate resource ownership. For some people coming from the Lua or C world, making Lua do more may be easier to accept. However, in a real-world project, these details will be hidden. As seen in this section, if you have an abstraction, it is easy to change from one way to another.

Next, I would like to introduce you to the RAII resource management idiom.

What is RAII?

This chapter is all about resource management. A resource can be a piece of memory, an opened file, or a network socket. Although in this chapter we have only used memory management as examples, the principles for all resources are the same.

Of course, all acquired resources need to be released. In C++, the destructor is a good place to release resources. When working with Lua, the Lua finalizer is a good trigger to release resources.

Resource Acquisition is Initialization, or **RAII**, is a useful resource management idiom. This means object creation and acquiring the resources the object needs should be an atomic operation – everything should succeed, or the partially acquired resources should be released before raising an error.

By using this technique, the resources are also linked with the object's life cycle. This ensures that all resources are guaranteed to be available during the life cycle of the object. This will prevent complex failure scenarios. For example, say a job has been half done and resources were spent but it could not be finished due to a certain new resource not being available.

When designing a C++ class, you can make sure all resources are acquired in the constructor and released in the destructor. When integrating with Lua, make sure you provide a finalizer and destroy the object from the finalizer.

The finalizer will be called during one of Lua's garbage collection cycles. You should not assume when Lua does this as it is not portable across different platforms and Lua versions. If you are memory-constrained, you can trigger a garbage collection cycle manually by calling `lua_gc` from C++ or `collectgarbage` from Lua.

> **Garbage collection is only for memory**
>
> Remember that garbage collection is only for memory. If you have a simple class that does not use other resources, it may be tempting to not provide a Lua finalizer. But if later the class is changed to rely on a non-memory resource, chances are that adding a finalizer is not part of the change. Then, you find out later that the resource is leaked from a weird bug report.

RAII is also useful in multithreaded programming to acquire a shared resource. We will see an example of this in the next chapter.

Summary

In this chapter, we learned more about resource management. We learned how to provide a customized memory allocation function to Lua. We also learned how to hold the actual C++ object in a Lua userdata. Finally, we familiarized ourselves with the RAII resource management technique.

In the next chapter, we will explore multithreading when integrating Lua into C++.

11

Multithreading with Lua

In the previous chapter, we learned some techniques to manage resources when integrating Lua into C++. In this chapter, we will learn how to work with multithreading with Lua. If you use Lua in a complex project, chances are that you need to create multiple Lua instances. First, we will learn how to contain the multithreading part in C++ and make Lua unaware of it. Then, we will see how Lua handles multithreading, in case you need to use it. Understanding multithreading will help with the technical planning for your projects.

We will cover the following topics:

- Multithreading in C++
- Multithreading in Lua
- Using `coroutine` with C++

Technical requirements

We will use the source code from *Chapter 10* as a base to develop the examples in this chapter. Make sure you can access the source code for this book: `https://github.com/PacktPublishing/Integrate-Lua-with-CPP/tree/main/Chapter11`.

Multithreading in C++

What is **multithreading**?

There are a few definitions, depending on the point of view. From the CPU's perspective, a multi-core processor that can execute multiple threads of instructions concurrently is real multithreading. From an application's perspective, using multiple threads is multithreading. From a developer's perspective, more focus might be on thread safety and various synchronization mechanisms, which are not multithreading itself, but its implications.

In this section, we will learn how to use Lua with C++'s native multithreading support. Each C++ thread will have its own Lua state. Because the Lua library does not keep any state and Lua states are not shared among different threads, this is thread-safe.

How does C++ support multithreading?

Since C++11, the standard library supports multithreading with `std::thread`. Each `std::thread` instance represents a thread of execution. The most important thing to provide to a thread is the thread function. This is what a thread executes. In its simplest form, we can create a thread as follows:

```
void threadFunc(...) {}
std::thread t(threadFunc, ...);
```

Here, we passed a C++ function as the thread function to create a thread. The function can optionally take arguments and the `std::thread` constructor will forward the arguments to the thread function. After the thread is created, the thread function starts to execute in its own thread. When the thread function finishes, the thread is ended.

We can also use a class member function or a class static member function as the thread function by invoking different constructors. You can refer to a C++ reference manual to learn more about `std::thread`.

> **Before C++11**
>
> In the era before C++11, there was no standard multithreading support. People had to use third-party libraries or implement their own with a low-level library, such as **pthreads**.

This type of multithreading is unlikely to surprise you. This is the type of multithreading that people have talked about and have used most, which is **preemptive multithreading**. The thread function can be paused at any time and resumed at any time.

Next, we will explore a real example to see C++ multithreading in action.

Using multiple Lua instances

In this section, we will implement a thread function in which we'll execute a Lua script. Then, we will create multiple threads to execute this same thread function.

Based on the source code from *Chapter 10*, wipe `main.cpp` clean and add the following code:

```
#include "LuaExecutor.h"
#include "LoggingLuaExecutorListener.h"
#include <iostream>
#include <mutex>
#include <thread>
#include <vector>

auto listener = std::make_unique
    <LoggingLuaExecutorListener>();
std::mutex coutMutex;
```

Here, we added the necessary headers. `listener` is the Lua executor listener and will be shared for all Lua executor instances. `coutMutex` is a mutex for printing results with `std::cout`, whose usage we will see next.

Next, implement the thread function, as follows:

```
void threadFunc(int threadNo, int a, int b, int c)
{
    auto lua = std::make_unique<LuaExecutor>(*listener);
    lua->execute("function add_params(a, b, c) return a + b
        + c end");
    auto result = lua->call("add_params",
        LuaNumber::make(a), LuaNumber::make(b),
        LuaNumber::make(c));

    std::lock_guard<std::mutex> lock(coutMutex);
    std::cout << "[Thread " << threadNo << "] "
              << a << "+" << b << "+" << c << "="
              << std::get<LuaNumber>(result).value
              << std::endl;
}
```

The thread function takes three integers as arguments, creates a Lua executor, and executes a Lua script to add the three integers. Then, it prints out the result.

Because there is only one place the standard output can print to, we are guarding the standard output with a mutex. Otherwise, the output sequence will be a mix of different threads and unreadable.

The way we use this mutex is by creating `std::lock_guard` instead of calling `std::mutex::lock` and `std::mutex::unlock` directly. The lock guard will acquire the mutex during construction and release the mutex when it goes out of scope and gets destroyed. This is an example of the *RAII* principle.

> **Recap of RAII**
>
> In the previous chapter, we learned about Resource Acquisition is Initialization (RAII). The C++ standard library adopts this principle in numerous places. Suppose that we do not use the lock this way, but acquire and release it manually. If anything goes wrong in between, there is a risk that the lock is not released in a thread and breaks the whole application. With the lock guard, the lock is always released even if an exception is raised because the C++ language guarantees that the lock's destructor is called when the lock goes out of scope. Before C++11, people would implement their own lock guard by creating a wrapper class that acquires the lock in the constructor and releases the lock in the destructor. This idiom is called **scoped locking**. C++17 also provides `std::scoped_lock`, which can lock on multiple mutexes.

Finally, let's implement the `main` function, as follows:

```
int main()
{
    std::vector<std::thread> threads;
    for (int i = 0; i < 5; i++)
    {
        int a = i * 3 + 1;
        threads.emplace_back(threadFunc, i + 1,
            a, a + 1, a + 2);
    }
    for (auto &t : threads)
    {
        t.join();
    }
    return 0;
}
```

This creates a list of threads and waits for the threads to finish execution.

In the first `for` loop, we use `std::vector::emplace_back` to create the threads at the end of the vector in place. Internally, for most C++ implementations, it uses *placement new* and invokes `std::thread(threadFunc, i, a, a + 1, a + 2)`. We do this because `std::thread` is not copy-constructible. Understandably, it does not make sense to copy a thread.

In the second `for` loop, we use `std::thread::join` to wait for all threads to finish execution. The `main` function runs in the main thread of the application process. When `main` exits, all other threads will be aborted, even if they have not finished execution.

Next, we'll test our example.

Testing it out

Compile and execute the project. You should see an output similar to the following:

```
Chapter11 % ./executable
[Thread 2]  4+5+6=15
[Thread 3]  7+8+9=24
[Thread 5]  13+14+15=42
[Thread 1]  1+2+3=6
[Thread 4]  10+11+12=33
```

If you run the project multiple times, you will see the order of the results from different threads changes. This verifies that we are using Lua with multiple threads.

For most projects, when integrating Lua into C++, this mechanism should suffice for multithreading. This is multithreading with C++. The Lua part just works without any additional effort. Each C++ thread has its own Lua instance and executes its copy of the Lua scripts. Different Lua instances do not interfere with or know about each other.

Next, we will explore multithreading in Lua.

Multithreading in Lua

To understand multithreading in Lua, let's begin with a fundamental question.

How does Lua support multithreading?

Lua does not support multithreading. Period.

But we cannot finish this section yet. We will explain this further with two approaches – a contemporary one and an old-school one.

The contemporary approach

Lua is a scripting language and it does not support *preemptive multithreading*. It simply does not provide a library function to create a new thread, so there is no way to do it.

Nowadays, CPUs and operating systems are designed around *preemptive multithreading* – that is, a thread of execution can be paused and resumed at any time. A thread has no control over its execution schedule.

However, Lua provides a mechanism for **cooperative multithreading** with **coroutines**. In a *cooperative multithreading* environment, the thread of execution is never preempted. Only when the thread willingly gives up its execution can another thread start to execute using the same CPU core. An application component that can be executed this way is called a `coroutine`, which is usually a function.

`coroutine` is also very popular with Kotlin for Android and backend development.

> **Cooperative multithreading**
>
> When we talk about threads, most of the time, the implication is that they are threads for CPU cores to execute. When we talk about *cooperative multithreading*, in some cases, such as the one for Lua, you may find that there is only one thread being executed and one CPU core used, even with coroutines. Arguably, this is not multithreading at all. But we do not need to judge. We need to understand this because multiple terms can be used for this in different contexts. We can also call this **cooperative multitasking**, which is technically more accurate from a historical point of view.

Let's see Lua's `coroutine` in action and explain it more.

Implementing a Lua coroutine

Replace the content of `script.lua` with the following code:

```
function create_square_seq_coroutine(n)
    return coroutine.create(function ()
        for i = 1, n do
            coroutine.yield(i * i)
        end
    end)
end
```

`create_square_seq_coroutine` creates a `coroutine` with `coroutine.create`, which, in turn, takes an anonymous function as its argument. You can roughly think that the inner anonymous function is `coroutine`. The inner function runs a loop and **yield** the squares from 1 to n.

You can only use `yield` with coroutines. A coroutine will stop execution when it reaches a `yield` statement. The values provided to `yield` will be returned to the call site, similar to what `return` does. The next time you execute `coroutine`, it will resume the execution from where it yielded until it reaches another `yield` statement or a `return` statement.

Let's start an interactive Lua interpreter to test our `coroutine`:

```
Chapter11 % ../lua/src/lua
Lua 5.4.6  Copyright (C) 1994-2023 Lua.org, PUC-Rio
> dofile("script.lua")
> co = create_square_seq_coroutine(3)
> coroutine.resume(co)
true    1
> coroutine.resume(co)
true    4
> coroutine.resume(co)
true    9
> coroutine.resume(co)
```

```
true
> coroutine.resume(co)
false    cannot resume dead coroutine
```

Here, we create a `coroutine` to return the squares from 1 to 3. The first time we `resume` `coroutine`, it starts to execute from the beginning and returns two values, `true` and `1`. `true` is from `coroutine.resume` and means that `coroutine` is executed without any error. `1` is what `coroutine` yielded. The next time we `resume` `coroutine`, the loop continues with the next iteration and returns 4. Pay special attention to the line when `coroutine.resume` only returns one value. The loop has finished but there is still code to be executed for `coroutine`, such as the implicit return statement. So, `coroutine.resume` returns `true`. After that, `coroutine` has finished and cannot be resumed and `coroutine.resume` will return `false` with an error message.

If this is the first time you have used `coroutine` with any programming language, this may seem magical and non-logical to you. How could a function, not in a thread, not reach its end and get executed from the middle of it again? I will explain why this is so ordinary (but do say you know `coroutine` and why it is so glorious in an interview) in the last part of this section. Before that, let's explore another example to see a case in which `coroutine` can be very useful.

Lua coroutine as iterator

We have seen how to use iterators with the *generic for* to simplify our lives, for example, `ipairs`.

But what is an iterator?

An **iterator** is something that can be called again and again to produce values until there are no more to produce. For Lua, `iterator` returns an **iterator function** that can be called again and again until it returns nil or nothing.

Based on `coroutine` that we have just implemented to generate a sequence of squares, let's build an iterator. In `script.lua`, add another function, as follows:

```
function square_seq(n)
    local co = create_square_seq_coroutine(n)
    return function()
        local code, value = coroutine.resume(co)
        return value
    end
end
```

`square_seq` is a Lua `iterator` as it returns its inner function as an `iterator function`. The inner function continues to resume the coroutine created with `create_square_seq_coroutine`. It is the caller's responsibility to stop calling the `iterator function` when the `iterator function` has returned nil or nothing.

Let's test this `iterator` in an interactive Lua interpreter:

```
Chapter11 % ../lua/src/lua
Lua 5.4.6  Copyright (C) 1994-2023 Lua.org, PUC-Rio
> dofile("script.lua")
> for v in square_seq(3) do print(v) end
1
4
9
```

You can see that three values are printed as expected for 1, 2, and 3.

And by looking at the usage, you cannot even tell if any coroutine or cooperative multithreading is involved. This, I think, is one of the examples where this programming paradigm can be more valuable than preemptive multithreading.

So far, we have explored Lua `coroutine` and Lua `iterator`. They can be more complex, but these examples are enough to show you how they work. You can refer to the Lua reference manual to learn more about `coroutine` and `iterator`.

Next, let me indulge myself by explaining this in my own terms.

Introducing multi-stacking

Traditionally, a thread is an execution unit for a CPU core, with its associated execution stack and the **program counter (PC)**. The PC is a CPU register for the address of the next instruction to be executed. As you can see, this is quite low-level and involves more details that we are not going to talk about.

Because this traditional image has been imprinted in us too much, even implicitly, it may have become an obstacle for you to understand coroutines.

Alternatively, let's seek help with one of the fundamental principles in computer science – **decoupling**.

The widely understood preemptive multithreading mechanism is already an application of decoupling. It decouples the thread from the CPU core. With it, you can have unlimited threads in the pool while you have limited physical CPU cores.

When you accept this, we only need to go one step further. If you accept that the execution stack can be decoupled from the execution thread as well, that is how `coroutine` works.

In this cooperative multithreading mechanism, a thread can have its own pool of execution stacks. An execution stack contains the call stack and PC. So, now, we have a three-tier system.

I call these coroutines multi-stacking, a term I coined to better explain it. Have a look at *Figure 11.1*, which implicates the following:

- **The relationship between the CPU and threads**: There are more threads than CPU cores. A CPU core can execute any thread. When a thread is resumed, it can be picked up by any CPU core. This is the *preemptive multithreading* that we know of, which usually requires CPU hardware support and is managed by the operating system transparently.

- **The relationship between threads and coroutines**: One thread can have multiple coroutines, each with its own execution stack. Because the operating system stops at the thread level, the operating system has no concept of coroutines. For a thread to execute another coroutine, the current coroutine must give up its execution and yield willingly. This is why it is called *cooperative multithreading*. The coroutine has no concept of threads either; its owning thread can be preempted and picked up by another CPU core later, but these are all transparent to the coroutine. A coroutine can also be picked up by different threads if the programming environment supports that.

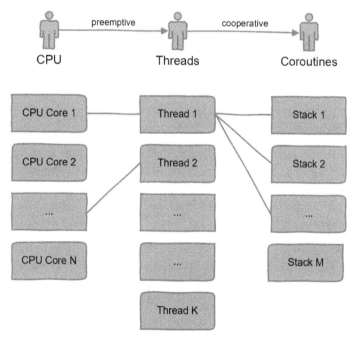

Figure 11.1 – Multi-stacking

Take a moment and think about the two relationships, which are explained and illustrated in *Figure 11.1*. This is one of the ways, albeit be unusual, to explain coroutines. The goal here is to find similarities between different mechanisms and technologies.

Next, we will look at cooperative multithreading and `coroutine` from another perspective.

The old-school approach

So far in this section, we have focused on the contemporary approach to explain the `coroutine` concept. Hopefully, with the two relationships explained, you can understand and tell why `coroutine` adds another valuable layer of multithreading support in a modern computing system.

However, this is nothing new. This is how computers have worked since the beginning.

Of course, in the very beginning, as you know, we feed the machine paper tape with holes. So, it is hopeless for multithreading.

Then, later, it becomes more complex. But still, there is only one processing unit in the CPU, and there is a privileged and primitive control program running in a dead loop. The main thing the loop does is check if another program wants to run. If there is one, it loads the starting address of that program into the PC – the program counter of the CPU. Then, the CPU starts to execute the other program.

You have probably guessed the problem. What if there is yet another program that wants to run? As it used to be, it has to wait for the first program to finish. This, as you can imagine, is not fair. So, we improve it and regulate that all programs should play nicely and yield their execution once in a while. Doing this allows the privileged control program to resume and find out if another program needs to run.

This is called *cooperative multithreading*. Each of the programs, besides the privileged one, is a `coroutine` instance, except that this term had not been invented in that era.

It helps, but not always. Suppose that one program decides to wait for an I/O that never happens and does not yield; the computer will be waiting endlessly in vain.

Much later, the computer became more powerful and could support running more complex operating systems. It moved the logic to decide which program to run into the operating system. If a program is waiting for I/O or has been running for enough time, the operating system will pause it and resume another program to run. This is *preemptive multithreading*. As it turns out, this is the right move. The operating system is fairer and the computer can do more.

Fast-forward to recent years – Moore's law no longer applies or at least has been paused. So, the CPU is not getting 1,000 cores, but the threads in a working computer are ever-increasing. Thus, the cost for the operating system to preempt and iterate through all the threads has now become a concern.

What can we do?

Some smart guys found out that we just need to do what we did in the beginning – use cooperative multithreading again. But this time, the controlling program is your main program – since you can't be selfish with yourself, you will be fair to all your coroutines to the best of your ability.

This is a simplified version of the evolution of the computer system. It is not historically perfect and has some dramatic touches to it. The goal is for you to realize that `coroutine` is a simple concept and that you can be comfortable with it.

Next, we will learn how to use coroutines with C++.

Using coroutine with C++

Do not use Lua `coroutine` with C++ if you have an alternative. As the iterator example showed, you can wrap `coroutine` in a normal function and keep calling it until it returns nil.

But to be less opinionated and for completeness, we can use the following Lua library function to start or resume a coroutine from C++:

```
int lua_resume (lua_State *L, lua_State *from, int narg,
                int *nresults);
```

The function is similar to `pcall`. It expects the function to be called and, optionally, its arguments on the stack. The function will be the coroutine. `L` is a stack for the coroutine. `from` is the stack from which the coroutine is called. `narg` is the number of arguments to the `coroutine` function. `nresults` points to an integer and Lua will output the number of values yielded or returned to the integer.

Let's see an example to understand how this works. In `LuaExecutor.h`, add a function declaration, as follows:

```
class LuaExecutor
{
public:
    LuaValue resume(const std::string &function);
};
```

The `resume` function supports `coroutine` in a limited way. It does not support passing parameters, but this is easy to do and you can refer to `LuaExecutor::call`. It only expects one return value as an integer. This is just to demonstrate the Lua API, not to add complete support for `coroutine` in our Lua executor.

Implement `resume`, as follows:

```
inline LuaValue
LuaExecutor::resume(const std::string &function)
{
    lua_State *thd = lua_newthread(L);

    int type = lua_getglobal(thd, function.c_str());
    assert(LUA_TFUNCTION == type);

    int nresults = 0;
    const int status = lua_resume(thd, L, 0, &nresults);

    if (status != LUA_OK && status != LUA_YIELD)
    {
        lua_pop(thd, 1);
```

```
        return LuaNil::make();
    }
    if (nresults == 0)
    {
        return LuaNil::make();
    }

    const int value = lua_tointeger(thd, -1);
    lua_pop(thd, nresults);
    lua_pop(L, 1);
    return LuaNumber::make(value);
}
```

This is a five-step process, separated by newlines, which is explained as follows:

1. It creates a new Lua state with `lua_newthread`. A reference to the new state is also pushed onto the main stack owned by L. We can call this new state the coroutine state. But `thd` is the coroutine stack. The Lua `lua_newthread` library function creates a new Lua state that shares the same global environment with the main state L but has its own execution stack. Yes, the API name is a bit misleading, but it is what it is.

2. It pushes the function name to be executed as a coroutine onto the new Lua stack.

3. It calls `lua_resume` to start or resume the coroutine. Since we are always creating a new state named `thd`, we are always starting the coroutine afresh. To resume it, we need to save `thd` somewhere and pass it in for future calls.

4. It checks whether there is an error; or, in case the coroutine does not return any result, it means that it has ended.

5. It retrieves the single integer that we expect, pops it from the coroutine stack, pops the reference to the coroutine stack from the main stack, and returns the value.

> **Taking care of other Lua states**
>
> A coroutine needs its own Lua state to be executed. You need to keep a reference to its Lua state somewhere until you no longer need the coroutine. Without a reference, Lua will destroy the state during garbage collection. If you have many coroutines, keeping all of those extra Lua states in the main stack can be messy. So, if you want to work with coroutines in C++, you need to design a system to hold and query those Lua states.

Next, add the following Lua function to `script.lua`:

```
function squares()
    for i = 2, 3 do
        coroutine.yield(i * i)
```

```
      end
  end
```

This function yields two values. The `for` loop is hardcoded because we do not support passing arguments in `LuaExecutor::resume`.

For the last bit of the demonstration, write `main.cpp` like so:

```cpp
#include "LuaExecutor.h"
#include "LoggingLuaExecutorListener.h"
#include <iostream>

int main()
{
    auto listener = std::make_unique<
        LoggingLuaExecutorListener>();
    auto lua = std::make_unique<LuaExecutor>(*listener);
    lua->executeFile("script.lua");
    auto result = lua->resume("squares");
    if (getLuaType(result) == LuaType::number)
    {
    std::cout << "Coroutine yields "
            << std::get<LuaNumber>(result).value
            << std::endl;
    }
    return 0;
}
```

This sets up the Lua executor and calls the `resume` function to start `coroutine`.

Compile and run the project; you should see the following output:

```
Coroutine yields 4
```

This shows how to work with `lua_resume`. You can read the Lua reference manual to get more detailed information about this API.

C++ code can also be executed as a coroutine. This can be done where a `lua_CFunction` implementation is provided to `lua_resume`, or Lua code in a coroutine calls a `lua_CFunction` implementation. In this case, C++ code can also yield by calling `lua_yieldk`.

Using coroutine with C++ can be very complex, but if you have your use cases defined, this can be abstracted to hide the complex details. This section is only an eye-opener. You can decide whether to use Lua this way or not.

Summary

With that, we have wrapped up the final chapter of this book. In this chapter, we focused on multithreading mechanisms, preemptive multithreading and cooperative multithreading, and Lua coroutines.

Lua coroutines can be used without C++ for advanced Lua programming and you can hide all these details from C++. We only touched the tip of the iceberg. You can read the Lua reference manual and practice more. You can also explore more on how to use coroutines with C++ by experimenting with the related Lua library functions.

In this book, we implemented `LuaExecutor` progressively. Each chapter added more features to it. However, it is not perfect. For example, `LuaValue` can be improved to make it easier to work with, and `LuaExecutor` can support more table operations. You can use `LuaExecutor` as a base and adapt it to your project or implement your own in a completely different way after you have learned the mechanisms.

I am confident that at this point, you can make improvements and add more features that suit you the best. You can always revisit the chapters as reminders and search the Lua reference manual for what you need.

Index

Packtpub.com

Subscribe to our online digital library for full access to over 7,000 books and videos, as well as industry leading tools to help you plan your personal development and advance your career. For more information, please visit our website.

Why subscribe?

- Spend less time learning and more time coding with practical eBooks and Videos from over 4,000 industry professionals

- Improve your learning with Skill Plans built especially for you

- Get a free eBook or video every month

- Fully searchable for easy access to vital information

- Copy and paste, print, and bookmark content

Did you know that Packt offers eBook versions of every book published, with PDF and ePub files available? You can upgrade to the eBook version at packtpub.com and as a print book customer, you are entitled to a discount on the eBook copy. Get in touch with us at customercare@packtpub.com for more details.

At www.packtpub.com, you can also read a collection of free technical articles, sign up for a range of free newsletters, and receive exclusive discounts and offers on Packt books and eBooks.

Other Books You May Enjoy

If you enjoyed this book, you may be interested in these other books by Packt:

Modern C++ Programming Cookbook - Second Edition

Marius Bancila

ISBN: 978-1-80020-898-8

- Understand the new C++20 language and library features and the problems they solve
- Become skilled at using the standard support for threading and concurrency for daily tasks
- Leverage the standard library and work with containers, algorithms, and iterators
- Solve text searching and replacement problems using regular expressions
- Work with different types of strings and learn the various aspects of compilation
- Take advantage of the file system library to work with files and directories
- Implement various useful patterns and idioms
- Explore the widely used testing frameworks for C++

C++ High Performance - Second Edition

Björn Andrist, Viktor Sehr

ISBN: 978-1-83921-654-1

- Write specialized data structures for performance-critical code
- Use modern metaprogramming techniques to reduce runtime calculations
- Achieve efficient memory management using custom memory allocators
- Reduce boilerplate code using reflection techniques
- Reap the benefits of lock-free concurrent programming
- Gain insights into subtle optimizations used by standard library algorithms
- Compose algorithms using ranges library
- Develop the ability to apply metaprogramming aspects such as constexpr, constraints, and concepts
- Implement lazy generators and asynchronous tasks using C++20 coroutines

Lua Quick Start Guide

Gabor Szauer

ISBN: 978-1-78934-322-9

- Understand the basics of programming the Lua language
- Understand how to use tables, the data structure that makes Lua so powerful
- Understand object-oriented programming in Lua using metatables
- Understand standard LUA libraries for math, file io, and more
- Manipulate string data using Lua
- Understand how to debug Lua applications quickly and efficiently
- Understand how to embed Lua into applications with the Lua C API

Packt is searching for authors like you

If you're interested in becoming an author for Packt, please visit `authors.packtpub.com` and apply today. We have worked with thousands of developers and tech professionals, just like you, to help them share their insight with the global tech community. You can make a general application, apply for a specific hot topic that we are recruiting an author for, or submit your own idea.

Share Your Thoughts

Now you've finished *Integrate Lua with C++*, we'd love to hear your thoughts! Scan the QR code below to go straight to the Amazon review page for this book and share your feedback or leave a review on the site that you purchased it from.

`https://packt.link/r/1805128612`

Your review is important to us and the tech community and will help us make sure we're delivering excellent quality content.

Download a free PDF copy of this book

Thanks for purchasing this book!

Do you like to read on the go but are unable to carry your print books everywhere?

Is your eBook purchase not compatible with the device of your choice?

Don't worry, now with every Packt book you get a DRM-free PDF version of that book at no cost.

Read anywhere, any place, on any device. Search, copy, and paste code from your favorite technical books directly into your application.

The perks don't stop there, you can get exclusive access to discounts, newsletters, and great free content in your inbox daily

Follow these simple steps to get the benefits:

1. Scan the QR code or visit the link below

https://packt.link/free-ebook/9781805128618

2. Submit your proof of purchase
3. That's it! We'll send your free PDF and other benefits to your email directly